GIZA

THE TESLA CONNECTION

"Is the Great Pyramid of Giza just a 'tomb and tomb only,' or is there much more to the last surviving wonder of the ancient world than the limited minds working within the cult of Egyptology are capable of grasping? There have been many theories but Chris Dunn's latest book, *Giza: The Tesla Connection*, is by far the most important, detailed, and convincing case ever made for a lost advanced technology of prehistory that left its mark on the sublime works of the ancient Egyptians."

GRAHAM HANCOCK, AUTHOR OF *FINGERPRINTS OF THE GODS*,
MAGICIANS OF THE GODS, AND *AMERICA BEFORE*

"This is masterful work. *Giza: The Tesla Connection* provides a comprehensive, engineering-focused hypothesis for one of the world's greatest mysteries. Accounting for the myriad of recent new discoveries surrounding the Great Pyramid, Dunn ties these together with disparate scientific advances into a cohesive and comprehensive explanation for its purpose and function, resulting in mind-bending conclusions and staggering implications for our understanding of the history of civilizations on planet Earth. Dunn playfully entertains speculation and takes the reader on a journey of technological wonder into the unknown past and future with appropriate context and an open-minded approach, something sadly lacking from modern-day academic Egyptology. As the inevitable progression of our scientific understanding of ancient technology slowly forces a reimagining of our own past, Dunn's body of work will be recognized for the groundbreaking strides it truly represents in unraveling the mysteries of the ancient Egyptian's, and our, ancestors. Like the work of his engineering-focused predecessor Sir Flinders Petrie, this book is required reading for anyone wishing to delve into these fascinating topics."

BEN VAN KERKWYK, AUTHOR AND
CONTENT CREATOR AT UnchartedX.com

"Dunn's shocking and cutting-edge research on nonpolluting free energy is forcing Egyptologists to think the unthinkable—that the technology on the Giza Plateau was more advanced than our current science. Bravo, Chris! A must-read for deep ecologists and readers seeking the real story of the past. At last! We have the answer to the riddle of the Great Pyramid."

BARBARA HAND CLOW, AUTHOR OF
AWAKENING THE PLANETARY MIND
AND *REVELATIONS FROM THE SOURCE*

GIZA

THE TESLA CONNECTION

Acoustical Science and
the Harvesting of Clean Energy

CHRISTOPHER DUNN

Bear & Company
Rochester, Vermont

Bear & Company
One Park Street
Rochester, Vermont 05767
www.BearandCompanyBooks.com

Text stock is SFI certified

Bear & Company is a division of Inner Traditions International

Cataloging-in-Publication Data for this title is available from the Library of Congress

ISBN 978-1-59143-461-0 (print)
ISBN 978-1-59143-462-7 (ebook)

Printed and bound in the United States by Lake Book Manufacturing, LLC
The text stock is SFI certified. The Sustainable Forestry Initiative® program
promotes sustainable forest management.

10 9 8 7 6 5 4 3 2 1

Text design by Kenleigh Manseau and layout by Debbie Glogover
This book was typeset in Garamond Premier Pro with Gill Sans MT Pro and S&S
Amberosa Sans used as display typefaces
Figures 1.1, 2.2, 2.3, 4.6, and 9.8 created by Jeff Summers, Media One Visual Arts.

To send correspondence to the author of this book, mail a first-class letter to the
author c/o Inner Traditions • Bear & Company, One Park Street, Rochester, VT
05767, and we will forward the communication, or contact the author directly at
gizapower.com.

To my wife, Jeanne Simpson Dunn.
Without her love, patience, protection, and guidance,
this book would not exist.

CONTENTS

▲

AN EGYPTIAN ENGINEER'S POINT OF VIEW

AHMED ADLY

Growing up in Egypt, I studied at school all the traditional theories about the "Pyramid Tomb" having been built with primitive tools and visited them a couple of times, all the while not focusing much attention on the science and knowledge behind these enigmatic treasures; this is a short story of Egyptians of my age. Time passed, and I worked as an electrical and systems engineer in a global telecom company in Giza. For years, I kept passing by the pyramids on my way to the office, looking at them and asking if all that effort was worth only hosting the king's body after his death?

Later in my life, I had the chance to read Chris Dunn's two books, *The Giza Power Plant* and *Lost Technologies of Ancient Egypt;* after this moment, things that had not made sense to me before began to make sense. The ideas proposed in his books are logical and striking at the same time. Despite many questions being answered, more questions were raised. It was a turning point for me, and I have begun to look at the monuments differently.

I decided to challenge some of Dunn's ideas on a deeper level, such as the Serapeum boxes' precision. In October 2019, I brought two digital measuring devices, a laser distance measure from Bosch, with an accuracy of 1/16 inch (1.5 mm) for 165 feet (50 m) and a digital protractor with an accuracy of +/− 0.3 degrees to measure the inside corners. *Digital instruments far surpass the precision of the measuring instruments used by the ancient Egyptians,* I told myself while standing in front of one of the giant boxes I had randomly chosen. Applying the instruments, I was astonished to find that the internal measurements are so precise, with unique characteristics of parallelism and right-angled corners of 90.0 degrees (not 90.1 and not 89.9). Measuring the width and length of the inside of the box from four corners, the lengths were both 3.091 meters (10.141 feet) and the widths were 1.495 meters (4.904 feet). In addition, when identifying what craft skills would be needed to produce artifacts with such precision during the modern era of industrial development, they would be skilled machinists and die makers. Not quarry workers, carpenters, or sculptors.

I came out of the underground tunnels telling myself, *It seems that Chris Dunn was right! Advanced methods must have been used to make these boxes.* While I had the permit to measure just one box, I later learned from other friends and colleagues who applied different measures inside the corners of other boxes, that some are 90° and others are found to be less than perfect conditions, though not varying enough to

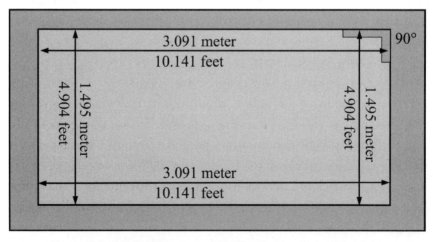

Figure F.1. Adly's measurements of a Serapeum granite box.
Image credit: Ahmed Adly.

ignore the significance of the artifacts. Additional analysis and unified measures are required to understand these boxes. However, in my opinion, applying only primitive methods to carve and transport these boxes is highly doubtful and still unverified.

Not only are modern visitors startled by the boxes' sizes but so was August Mariette, who in *Le Sérapéum de Memphis* wrote, "The dimensions of the granite sarcophagi found in the large undergrounds are well made to amaze the visitor. This leads me to record here a note on an interesting question of ancient mechanics, how the Egyptians could have introduced such masses into the depths of an underground passage and into chambers from which it would certainly be very difficult to draw them, even with the help of the most ingenious complications of modern mechanics" (Mariette 1882).

By the date of the discovery of the Serapeum in 1851, the machine era had already started a while ago; there were already locomotives, lifters, hydraulic presses, moving machinery, steam engines, and more. I guess that Mariette wasn't far in his assumption; however, he didn't find any kind of machinery in the tunnels except wooden winches, so he established the transport method of the boxes based on what he had found.

I filmed the measurement operation and published it in a video in Arabic (Adly 2021). I also applied different measures for several other artifacts, like obelisks, statues, and pillars. Sharing these observations in the form of magazine articles and videos gradually grabbed Egyptian and Arabic speakers' attention to the fact that we have missed a vital piece of ancient Egyptian history.

For a long time, the pyramids were a mystery, and they still are. There are always three areas of debate when discussing the pyramid tomb theory with professional Egyptologists, especially when talking about the Fourth Dynasty giant pyramids in Giza and Dahshur. Unambiguously, we are assured that they are tombs. But why are there no original mummies? Why the lack of inscriptions? Why does their interior architecture look nothing like ordinary tomb design, which is apparent in the ancient tombs in the Valley of the Kings in Luxor and other sites? Why are these things missing from the Fourth Dynasty pyramids?

Mainstream academics still claim that the reason for not finding any original mummies buried in pyramids is because they were robbed in antiquity. While this is a commonly held view, valid arguments can be made against it. In the early 1900s, Italian archaeologist Alessandro Barsanti explored the great pit of the Zawyet el Aryan pyramid. Under 65 feet (20 m) of large blocks, he found a sealed giant granite sarcophagus. On removing the lid they found it was empty, except for the black material that lined the side walls. Barsanti was confused as to why one would have taken the care of concealing the sarcophagus under an enormous mass of blocks if it had been empty (Barsanti 1906). This was not the only such case. Around 50 years later, in 1954, Egyptologist Mohammed Zakaria Ghoneim excavated the unfinished pyramid of Sekhemkhet (a Third Dynasty king) in Saqqara (Ghoneim 1956). The burial chamber lies at 236 feet (72 m) from the entrance, and an alabaster sarcophagus stands in the middle of the room with a vertically sliding door at one end. The door was shut and plastered, and hopes were raised to a high pitch. Shockingly, when the sarcophagus was opened, it was also empty and unused. A question arose from the Egyptologists themselves without any clear answer, "Had any king ever been buried here?"

These two cases disprove the argument that "pyramids were tombs, but the evidence to prove it was removed by tomb robbers," and support the contention that pyramids were not built to be used as royal tombs. However, the absence of necessary evidence is ignored by academics. The only mummy attributed to a pyramid owner has been discovered inside the black basalt sarcophagus in the pyramid of Merenre (a Sixth Dynasty king). This mummy is today in the Imhotep Museum at Saqqara. Nonetheless, there are strong doubts about its identification. The great anatomist of the Egyptian mummies, Grafton Elliot Smith, studied and considered it to be of a much later period, possibly the Eighteenth Dynasty. If the pyramids were tombs, they should be viewed as the biggest failures in history, because they have never protected their owners' bodies.

The lack of hieroglyphs in the Fourth Dynasty pyramids is still a weak spot of the pyramid tomb theory, especially if we recognize that Khufu's children, Nefertiabet, Khufukhaf I, Meresankh II, and

Meresankh III, have well-written and colorful inscriptions in their tombs at the Giza Plateau. If so, why was inscribing the nine neighboring pyramids of Giza skipped (Fakhry 1969; Lehner 1997)?

The interior design of the pyramids is another serious argument. In a discussion with Hazem Zaki, a ministry of tourism and antiquities inspector on the Giza Plateau, I was wondering how King Sneferu would have been buried in his red pyramid at Dahshur. The pyramid's entrance passage is too narrow even for a single person; it is 3 feet (0.91 m) in height and 4 feet (1.2 m) wide and slopes down at 27 degrees for 200 feet (61 m). Putting the king's body alone on a sled is the only possible scenario, although not appropriate for his majesty. Zaki respectfully agreed that the burial procedure would be a nightmare and that we don't know how such funeral ceremonies could take place inside giant pyramids.

The true purpose for why the pyramids were built is what matters. This point could solve many mysteries. Everything should have reasoning for why it exists or is done. Ancient Egyptians cut and moved millions of stones, some of them from quarries that are hundreds of miles away, in a long and regular process for a purpose that would serve hundreds or thousands, not just a single person.

For millennia, explorers and historians were overwhelmed by the size and perfection of the pyramids of Giza. Still, one question mystified every visitor: Why did the architects decide to build the three pyramids of Giza with large stones? The average weight of a stone block in the core masonry of the Great Pyramid is 2.6 tons, and some weigh up to 15 tons. In the second pyramid, it is even more; we have estimated as much as 60 tons in the base. Why did they do so? Unless they wanted to create trouble for themselves and make the project impossible, the architects knew well that using small stones would have been easier and faster in raising the superstructure. Yet they may cause the pyramid function to fail! Could the giant interlocking rocks be there to withstand the underground vibrations proposed by Dunn?

It is not an exaggeration to say the pyramid machine theory marks a paradigm shift, not just for technical professionals but also for Egyptologists. During my discussions with many Egyptologists and tour guides, I was told that they are convinced by many documentaries and writings (based on Dunn's hypothesis) that the pyramids are not royal tombs. Still, unfortunately, for many reasons they cannot say that in public.

In *The Giza Power Plant*, an engineer looks at the Great Pyramid and sees something different. Dunn highlights and solves many ancient puzzles hardcoded among the rocks centuries ago. He proposes that the Great Pyramid of Giza was a power plant of ancient Egypt and investigates each of its unique features while providing a convincing and scientific explanation. He opened the door for many to look differently at our distant past. If our modern civilization relies on energy, why should we not assume the same for the ancients? Instead of burning fuel, though, they depended on the power generated from an exceptional understanding of the physical properties of the rocks. What everyone I have talked to admires about Dunn's research is that he not only tells you the direct conclusion but also shows the steps he followed, the analysis he did, and the observations he made to achieve a particular result.

Inspired by the pyramid machine ideas, many Egyptian engineers, chemists, and scholars believe that the effort behind building these massive structures was a rational one where every component was critical to its successful function.

In June 2021, after I had published a one-hour video on my YouTube channel explaining the end-to-end idea of the pyramid machine based on *The Giza Power Plant* book with the pump version from John Cadman, I received an email and later a phone call from A. M. El Sherbini, PhD, professor of experimental laser-plasma physics at the Faculty of Science, Cairo University, showing interest in the idea and a kind invitation to share his thoughts about it.

Professor Sherbini is a physicist who has made a global contribution to physics, has a method registered in his name for laser spectroscopic analysis of materials, and is a specialist in hydrogen-alpha. Fortunately, he is not far from Egyptian antiquities. In 2015, he gave a lecture in the Egyptian Museum in Cairo entitled "Recent Application of Laser in Archaeology," helping archaeologists use technology in their day-to-day discoveries.

Figure F.2. Professor Ashraf El Sherbini (right) *and engineer Ahmed Adly* (left) *in Sherbini's laboratory at Cairo University. Image credit: Ahmed Adly.*

In December 2021, I met with Sherbini in his laboratory at Cairo University to discuss several points around ancient Egypt and the tools used to build the remarkable monuments. The idea of *The Giza Power Plant* was one of the topics on our agenda, with its evidence, strengths, and weaknesses, and how to solve the outstanding points with more measurements and analysis. Although Sherbini is a challenging physicist, he supports the power plant hypothesis. These are some of his words and notes:

- The idea of the pyramid machine is brilliant and very reasonable. It is impossible that the Egyptian builders made such a considerable effort for just tombs. The interior design, including shafts and chambers, looks like a machine and does not fit the tomb theory.

- The subterranean chamber pump that generates pulses, the Queen's Chamber chemical reactions to produce hydrogen, and then the resonating King's Chamber is a logical sequence and provides answers to the design.
- The mysteries of the pyramids, how they were built, and their function are too difficult to be solved solely by Egyptologists. The involvement of experts and specialists in physics, chemistry, engineering, geology, and other branches of science, and an organization's support, are required to bring this idea to an actual model.

Scientific criticism is essential to improve. Sherbini and I believe that investigation is required in some areas that were discussed with Dunn. For example, more quantitative acoustic experiments are needed in the King's Chamber to get a better understanding of its vibration properties, and more scientific proofs are needed for generating electromagnetic energy from the granite slabs. You will find further explanations for these points as well as more pieces of evidence about the power plant theory in this book, for which I'm glad to have been asked to provide this foreword.

Day after day, we have found that the Great Pyramid is far more mysterious than we have imagined and more of its secrets are revealed using modern technology. Since *The Giza Power Plant* was published in 1998, several missions have been launched. The Djedi Project team sent a robotic tunnel explorer into the shafts of the Queen's Chamber using a pinhole camera to explore what is behind the stone block at its end and capture photos of what might be hidden. While ScanPyramids missions succeeded in detecting a big void of around 100 feet to 130 feet (30 m to 40 m) just above the Grand Gallery, and another corridor 30 feet (9 m) long has been discovered close to the main entrance of the great pyramid, these discoveries raise the question among people, television shows, and articles, about what could be the main pyramid function. I am happy to see that we now realize the

"New Paradigm" period that Dunn discussed twenty-five years ago in his first book.

AHMED ADLY was born in Kuwait and grew up in Egypt. In August 2006, he graduated from Ain Shams University with a BCS degree in Electrical and Systems Engineering. He is currently employed by Accenture in Düsseldorf, Germany. In 2016, Adly took up the challenge put forward in *Lost Technologies of Ancient Egypt* for Egyptian engineers to begin examining their own artifacts, and he began to examine, measure, and discuss with his peers the subject of lost technologies and share his observations to a wide, interested audience across Arabic-speaking nations through his YouTube channel, titled Ahmed Adly.

ACKNOWLEDGMENTS

I feel privileged and honored to have this opportunity to shine a light on the wonderful people who have contributed to my being able to write this book. Above all, I must express my love and respect to my parents, Peter and Celia McHugh-Dunn, who brought me into this world and who nurtured and guided me in a way that resulted in me flying away to follow a path none of us could have imagined at the time. My love and gratitude also go to my brother, Bernard (Winifred) Dunn, and sisters, Celia (Bahgat) El Nadi, Pauline Dunn, and Angela (Michael) Anderson.

My children have also been a source of inspiration. As any parent knows, what we do is mostly for our kids in the hope that we can not only provide for them when we are alive but leave something of value that they can use long into the future. Peter, Alexander, and Geno—thank you for being in my life and for your loving support.

As *The Giza Power Plant* started to take shape over 20 years before it was published in 1998, this sequel has been in the making for over 20 years since. In that book I acknowledged the many people who had provided critical information, help, and advice through the years of its development. While some appear below among those whose contribution is specific to this sequel, I thank them all again.

This book would not exist without the valuable technical knowledge and input provided by Arlan Andrews, ScD, Robert Vawter, Eric Wilson, and Friedemann Freund, PhD. With sadness and condolences to his wife, Helen, I must also report that on September 17, 2023, my dear friend and research associate Robert Vawter unexpectedly passed away. As you will read in the book, Robert was responsible for contributing a significant amount of valuable information to this work, including an appendix on pyramid acoustics and bringing to my attention physicist Friedemann Freund's research into earthquake lights.

I am also particularly grateful for the editorial feedback I have received from Judd Peck, JD, Robert Vawter, and Greg Brown. Also, my good friend Gary Lickfett, now passed on, who reviewed the manuscript while in hospice and provided his own unique observations and suggestions. For the part they have played in its development, my thanks also go to the Board of Directors and employees at Danville Metal Stamping, Ahmed Adly, Adrian Lungan, Susan Alexjander, Dustin Carr, PhD, Hamada Anwar, JD, Hany Helal, PhD, Taha Rabeh, PhD, Galal Hassaan, PhD, Mohamed Ibrahim, Lauren Kurth, John and Robert Barta of Barta's Precision Granite, Jeff Summers, Ben Van Kerkwyk, Next Anyextee, Sohaila Hussein, Ihab Rashad, Colin Higgins, Craig Docherty, Gamal Elfouly, Matt Sibson, Andy Leskowitz, Toan Trinh, Mike Mahar, John Cadman, Mike Cadman, Ralph Ellis, Mark Foster, Gardner Peck, and Tom Neal.

Having longtime supporters who regularly provide positive enforcement is critical for any person, regardless of their pursuits. For their unfailing support and encouragement for many years, I would like to thank Graham and Santha Hancock, Yousef Awyan, Shahrzad Awyan, Paul and Ardith Keller, Chris Keller, John Heckler, Jim Cox, Carol Radford, Carol Nichols, Steve Garcia, Dr. Michael Fuesting, Carmen Miller, Elena Lewis and Cassie Lewis-Babbs, Mark and Sonnia Schroeder, Phil Langley, Tim Welch, Jason Rome, Doug Keenan, Marcus Allen, Duncan Roads, Doug Kenyon, David and Jennifer Childress, Brenda Bush, Peter Brooks, Cliff Dunning, and David Wesner.

I began writing this book in June 2021 with a deadline of March 2022. After contracting COVID in September of that year, progress came to a halt and the residual effects I suffered from the virus limited my productivity. For their patience and understanding during slipped deadlines, I would like to thank Inner Traditions and Bear & Company, a publisher that has kept my previous books in print for over 25 years, for allowing me more time to deliver the final manuscript. Special thanks go to Barbara Hand Clow, Jon Graham, Kelly Bowen, John Hays, Jeanie Levitan, Courtney B. Jenkins, Manzanita Carpenter, Ashley Kolesnik, Albo Sudekum, Debbie Glogover, Lesley Allen, and all others who have devoted time and talent to bring this book to market. I appreciate you more than you know.

Finally, I would like to acknowledge anyone and everyone who has made an effort to explain the innumerable mysteries of our past. Regardless of how they have expressed what their minds have conceived, they have added to an important conversation that, I am sure, will continue to be discussed far into the future.

INTRODUCTION

The Giza Power Plant introduced a revolutionary idea and process for creating electricity, while redefining the original purpose of the pyramids of Egypt. This expansion of the technology discussed in *The Giza Power Plant* redefines the term used for energy production and moves away from the term "power plant." Generally, this term evokes visions of heavy industrial activity with the burning of fossil fuels or splitting atoms to create steam, which pushes turbine blades that are attached to electric generators, feeding their output into the electric grid. *Giza: The Telsa Connection* describes the Great Pyramid as a solid state (no moving parts) Electron Harvester. It is designed to shake electrons loose from igneous rock in the Earth, collect them, focus them, and radiate them out into the atmosphere in vast quantities. Perhaps instead of smoke stacks dotting the landscape 100 years into the future, we will see brightly glowing pyramids in the distance—*that is if they are not covered by a dome.*

THE TESLA CONNECTION

Nikola Tesla was born in 1856 and died in 1943. In the course of his career, he was awarded 308 patents, with 112 of them being awarded in the United States alone. In 1884, he emigrated to the US and worked for Thomas Edison for a period of time, but left to eventually compete with him, and have his invention of alternating current (AC) based power generation and distribution systems accepted over Edison's direct current (DC) based power generation and distribution systems. After prevailing in this acrimonious tussle, Tesla sold his AC related patents to George Westinghouse, who was awarded the contract to build generators and transmission equipment for the Niagra Falls Power Company.

Prior to emigrating to the US, Tesla worked for the Continental Edison Company in Paris. In 1883, Tesla was working on assignment in Strasbourg, when, on his own time and expense, he constructed his first induction motor.

When I started to write this book, Tesla's work was not the most prominent subject on my mind. It was only after all the elements of the electron harvester system were put together that it became clear that as he himself had intimated, what he had accomplished in his lifetime had been inspired by a Universal core of knowledge.

> *My brain is only a receiver, in the Universe there is a core*
> *from which we obtain knowledge, strength and inspiration.*
> *I have not penetrated into the secrets of this core, but I*
> *know that it exists. (Tesla 1856–1943, 13)*
>
> NIKOLA TESLA

I would like to suggest that this same core of knowledge, strength, and inspiration did not weaken over the millennia, and also inspired the designers and builders of the Great Pyramid.

Clues to what the Tesla Connection is can be found in the artwork on color plate 6. Specifically, the inclusion of the dome. This is not there for show, or the result of whimsical artistic license, but a natural evolution of an idea that evolved from the results of NASA geophysicist Friedemann Freund's research into earthquake lights, and Tesla himself when he was experimenting with the wireless transmission of energy from his famous Wardenclyffe Tower installation.

But that is not all. *Giza: The Telsa Connection* identifies Tesla's thumbprint on the blueprint of the Great Pyramid in both the deepest and dankest location in the Great Pyramid—the Subterranean

Chamber—and the tip of the pyramidion on the top. As you will learn, both areas identify a critical piece of information that every inventor who wishes to secure a patent must discover before the patent officer receives the application—that is, "prior art." An invention seeking patent protection that relies on previously patented inventions in order to work must disclose all prior art that existed.

This presents an interesting conundrum. If I applied for a patent for the Pyramid Electron Harvester, could I argue in good conscience that the application was not relying on prior art, or inventions that existed previously? I don't think so. Without the Great Pyramid, there is no way I could have formulated the theory you will read in this book or have read in my previous books. My policy has always been one of openness. It is normal to seek patent protection for an original idea if you wish to be recognized and compensated for the idea. This is different. *Giza: The Telsa Connection* describes a means of farming electrons from the Earth, and I strongly believe, and the evidence proves, that the idea is not originally mine. It originated thousands of years ago in Egypt. Rather than exploring the legality of making a claim, I decided to write a book. Obviously, I could not produce a working model when I was in the early stages of reverse engineering and needed to attract more knowledge, intellect, and talent to add to the discussion.

While it is strongly suggested that the pyramid builders and Tesla were drawing from the same source of knowledge, does this mean that Tesla was inspired in his endeavors by studying the Great Pyramid, or did his endeavors lead him to see a connection? If he was, he didn't talk about or write about it. However, had Tesla been born in our era, when serious questions about the validity of the pyramids being tombs for the pharaohs proliferate openly in society, I have no doubt that he could have drawn more inspiration from the Great Pyramid. Would Tesla even have imagined that his genius would eventually be resurrected from his tomb to help sweep away centuries of misunderstanding about a different "tomb"? As I look forward, the Tesla connection becomes increasingly more prominent in its relationship to the Great Pyramid's intent, design, and function.

QUESTIONS

Judge a man by his questions rather than his answers.

VOLTAIRE

We cannot expect to come up with the right answers if we don't ask the right questions. In 1971, a significant groundbreaking book was published by Harper & Row, New York. In *Secrets of the Great Pyramid* the author, Peter Tompkins, began his Introduction by posing this profound and inspiring question:

> Does the Great Pyramid of Cheops enshrine a lost science? Was this last remaining of the Seven Wonders of the World, often described as the most sublime landmark in history, designed by mysterious architects who had a deeper knowledge of the secrets of this universe than those who followed them? (Tompkins 1971, xiii)

Even asking the wrong question can sometimes yield benefits, if not ultimate truth. I have found during my lifetime that more knowledge can be gained from questions that are asked than answers that are given. The way Tompkins framed his question says a lot. It makes you sit back and think, "Wait a minute. The Great Pyramid, Cheops's tomb, contains a lost science? Why ask this question?" The question implies that the author has evidence to consider an affirmative answer. Moreover, the second part of his question suggests that not only was the science available to the pyramid builders lost since they built the structure, but it has remained lost for millennia since. Furthermore, the science is not specific to just the pyramids, but extends beyond the planet.

The right question can change your life. In just one paragraph of carefully chosen words, from the day in September 1977 when I opened his book, Tompkins had framed and influenced my life going forward. By the time I had closed *Secrets of the Great Pyramid*, I had become convinced that the answer to his questions was a resounding yes! For the next 21 years, my response to his question was carried forward with conviction, leading to the 1998 publication of my book, *The Giza Power Plant: Technologies of Ancient Egypt,* and my conviction has only become stronger since. The "lost science" enshrined in the Great Pyramid, as it turns out, is far greater than I had imagined and described in my 1998 book. While the book was considered revolutionary and controversial then, discoveries made since, in different fields of research, provide corrections for some areas and amplification for others. While answers to many questions remain, the book you now hold in your hand represents the many twists and turns, along with fortuitous chance events, that bring us closer yet to providing more accurate answers to Tompkins's questions.

The Giza Power Plant has received much support over the years from readers of various backgrounds. There has also been significant criticism. Some of it blunt and harsh, and some more measured and objective. One of the most reasonable and entertaining skeptical comments regarding my work came from a Professor of History at the University of Wales, Aberystwyth.

In his book, *Shadow Pasts: History's Mysteries,* after introducing the concept of the Great Pyramid being a coupled oscillator, along with a brief description of its various parts, William D. Rubinstein writes: "To most readers, all this will surely sound as if it comes straight from the heart of Fruitcake Land" (Rubinstein 2007, 162). Perhaps Rubinstein had heard that when I was young my father taught me how to make fruitcakes to accompany the Sunday roast, and my growing skill at doing so had been deeply embedded in my DNA—thereby influencing my life going forward.

It is an aspect of human nature to place a pejorative label on an idea that is unusual and unfamiliar. However, the reality is that new ideas and a fertile imagination have fueled the development of technology and invention—even from times when these preposterous and futuristic ideas were imagined, and laughed at, and while there were no precedents and no developed infrastructure to create them. Many of the technologies we enjoy today were previously conceived in a very special "Fruitcake Land" that was populated by science fiction writers. From this happy place we learned about flights to the moon, Space Odysseys, handheld communication devices that let you see who you are talking to, and many other flights of fancy and wild imaginings. Famous science fiction writer Arthur C. Clarke, who with Stanley Kubrick cowrote the screenplay for the 1968 movie *2001: A Space Odyssey,* wrote what became known as "Clarke's Third Law:"

> *Any sufficiently advanced technology is indistinguishable from magic.*
>
> ARTHUR C. CLARKE

If we look at the technology that existed 100 years ago and how it was used, it is foreign and strange looking to most young people. Someone in their later years, like myself, who remembers what it was like to lower an arm with a sharp needle mounted in the end, into a tiny spiral groove cut into the surface of a vinyl disc, may sink into the mists of nostalgia until they are suddenly jolted into the twenty-first century by a tingly vibration in their pocket. Retrieving a black rectangular device, they swipe or press a region of the black screen and voice their usual greeting. They are now holding in their hands a small miracle of modern technology, in which they can store the music from every vinyl long-playing record they ever owned—and so much more. Sixty years ago, as they cleaned and buffed a vinyl disc before placing the center hole on a rod to lower it onto a platen, could they even have imagined a world in which they can access entertainment the way we do today? Would we give up what we now have and go back to the old technologies that we so fondly remember?

When I was being taught how to become a happy inhabitant of "Fruitcake Land," had I been able to see into the future and witness the inhabitants of planet Earth using their smartphones and talking to people across the planet, I would have thought it was magic. Even now, knowing something about how these devices were developed and the infrastructure that supports them, their pure genius is breathtaking.

IMAGINATION

When a distinguished but elderly scientist states that something is possible, he is almost certainly right. When he states that something is impossible, he is very probably wrong.

ARTHUR C. CLARKE (CLARKE 1964)

Imagination is the fuel used to bake the special fruitcakes that flow out of "Fruitcake Land." Failure of imagination would stifle creativity, excitement, and development and result in a society of humans who have stopped doing what they love to do. Whatever it is, we love to use our imaginations, see possibilities, and create. In spite of all naysayers and skeptics, humans push forward following their instincts and doing what they love.

Imagine then, if you will, being in a time machine and having the choice of being transported into the future or the past to witness activities on Earth in various cultures. You could trace the development of technologies from the beginning of the Industrial Revolution up to the present day. How communication devices have evolved, how transportation has changed. The electronics industry and the changes to television and radio. The creation and development of personal computers, and most startling, specifically in the electronics industry, the increase in functionality of the devices while at the same time how the cost of purchasing them decreases over time.

In 1974, I purchased a 27" Zenith television for $750. In 2022, I can purchase an internet-ready 27" flat screen TV, with greater picture quality than I could have imagined in 1974, for $150, and have it delivered to my door in two days. These advances can be traced to a quantum leap in progress in the electronics industry after the invention of a unique device in 1947. On December 23 of that year, William Shockley, John Bardeen, and Walter Brattain demonstrated the transistor at Bell Labs in Murray Hill, New Jersey.

Since then, transistors have made their way into most products that require electricity to operate. In 1972 they became a critical component in Liquid Crystal Displays (LCD) when T. Peter Brody and his team at Westinghouse incorporated thin film transistors in the first Active Matrix (AM) LCD. It didn't take many years after this until LCD screens replaced Cathode Ray Tube (CRT) screens on all computer monitors and television products, and as time passed, they began to appear on many other products too—even refrigerators.

As we sit in our time machine, we can't help but be tempted to go forward in time to see in what direction these technologies go and how far they reach. What present day technologies will be replaced by transformative technologies developed in the future? Will the transistor eventually become obsolete? Will it be replaced by an unforeseen/unanticipated discovery or invention that defies our current understanding of physics, one that will deliver greater performance for less cost? What limits are there to invention?

In 1902, Charles Holland Duell, who served as the United States Commissioner of Patents from 1898 to 1901, said, "In my opinion, all previous advances in the various lines of invention will appear totally insignificant when compared with those which the present century will witness. I almost wish that I might live my life over again to see the wonders which are at the threshold." Duell is also credited, by legend, with having said, *"Everything that can be invented has been invented."* This attribution has been dismissed for lack of evidence and is possibly linked to a joke that was published in *Punch,* a British comedy magazine. Duell died on January 29, 1920, at the age of 69 (Wikipedia).

How true do his prophetic words ring today when we can look back and see what wonders *were* at the threshold when he died?

What about other technologies? Will future inventions and developments see humans and their technologies settling on another planet? Considering that there were many advances in science and engineering born of the endeavor to reach the moon in the 1960s, I would say without hesitation that before a colony has been established on the Moon, or Mars, there will be significant advances in applications of technology along the way. When reflecting on what might exist 50 years in the future, it becomes mind-boggling to consider advances in technologies if they continue at the same pace they have had in the past 50 years.

In considering these possibilities, we might want to add into the mix a tantalizing and challenging demonstration of flight technology that may exist in the future, after suitable discoveries and invention take place. *This technology is already dancing before our eyes today.*

On June 25, 2021, a report was issued by the Office of the Director of National Intelligence (ODNI) that accepted the existence of UAPs (unidentified aerial phenomena), also referred to as UFOs (Unidentified Flying Objects), in US air space, and felt the evidence for their existence warrants further investigation. While dismissing many UAPs as natural phenomena or optical aberrations, the report admitted that some UAPs are genuine and exhibit abilities such as accelerating at great speed and sharply changing direction without slowing down. The capabilities of these craft far surpass what US technology is capable of, and presumably that of other countries as well, though the ODNI stops short of suggesting that they may have extraterrestrial origins.

> The US Navy recently admitted that, indeed, strangely behaving objects caught on video by jet pilots over the years are genuine head-scratchers. There are eyewitness accounts not only from pilots, but from radar operators and technicians.
>
> In August, the Navy established an Unidentified Aerial Phenomena (UAP) Task Force to investigate the nature and origin of these odd sightings and determine if they could potentially pose a threat to US national security.
>
> The recently observed UAPs purportedly have accelerations that range from almost 100 Gs to thousands of Gs—far higher than a human pilot could survive. There's no air disturbance visible. They don't produce sonic booms. These and other oddities have captured the attention of "I told you so, they're here!" UFO believers.
>
> But there's also a rising call for this phenomenon to be studied scientifically—even using satellites to be on the lookout for possible future UAP events. (ODNI 2021).

Without a clear answer as to where the advanced civilization that developed these craft is located, observing their performance makes one wonder what other advances would be apparent in the place where these machines were built. If we had the ability to be transported to that place where the inhabitants had been developing their technologies for perhaps thousands of years, not just hundreds as we have here on Earth, what would those technologies look like? By observing from a distance, what could we learn about them? Would we be like a child in the '50s peering into the future and seeing everybody using smart phones? It would probably all seem like magic to us. What would their cities look like? Equally important, what would the infrastructure needed to support this civilization look like? One would think that incredible knowledge and advances would feed into and be evident in all aspects of power generation, production, transport and services to maintain life.

Observing how these UAPs function leads one to conclude that the craft does not rely on burning fossil fuel to operate. They seem to be surrounded by some kind of corona or energy field, and there were no observations of fire bursting out of jet engine afterburners when a UAP was seen to accelerate at great speed. The UAP appeared to have total control over gravity and navigation. I can't help but think that if we were able to visit the place that created the craft, we'd find other examples of technologies that we don't have.

- Would we recognize the power plants that allow their civilization to operate?
- Does the civilization that created that craft use steam to rotate turbines to create electricity, or burn refined oil to power their vehicles the way we do?
- Outside their cities, at a comfortable distance away, would we see long vertical structures with smoke billowing out of the top?
- Would we see large engines on iron rails pulling hundreds of cars full of fossil fuels, such as oil and coal, destined to be burned to create the steam to drive turbines?
- Would we find oil refineries converting oil into gasoline to fuel their vehicles, or factories that convert oil into rubber that is used to make wheels for vehicles that roll along millions of miles of paved roads?

Surely if the UAP we have witnessed is any indication, we would probably see similar futuristic craft that magically whiz by at various elevations above the ground with barely a whisper of sound.

While the implied advanced technology demonstrated by UAPs may be a hundred or even a thousand years ahead of ours (though it is probably a mistake to characterize advancement as a reflection of years), it shows clearly that they have already mastered and can control the effects of gravity and inertia on physical objects and have an efficient propulsion system. The combination of both allows them to hover, accelerate at tremendous speed, make a 90-degree sharp turn, and descend from 80,000 feet to the surface of the Earth in a few seconds. On a craft built today, these movements would result in g-forces that would destroy the craft as well as the humans inside. It seems unlikely that a civilization that had hundreds of years of technological development would be relying on the same technology to fuel their society that was developed when they were taking their first steps along the path of technological progress.

FULL STEAM AHEAD

Here on planet Earth, the fundamental physics, science, and invention that allow us to travel, power our magical devices, and many other applications, have stayed basically the same. While quantum leaps in progress were being made in controlling how electrons behave within a device, the principal methods of power generation that harness and provide the electrons for the device have stayed fundamentally the same for over 100 years.

Excepting hydropower, power plants and engines still rely on motivation provided by a buildup of pressure from a gaseous medium. For coal, oil, gas and nuclear-fired power plants, it is steam, and for internal combustion engines, it is the explosion of gasoline or diesel fuel in an enclosed cylinder.

Taking all of this into account, it would seem that the infrastructure we have today, innovatively, has not kept pace with the rapid development of new technologies and the explosive growth of consumer use.

In the past, as consumer products were invented and introduced into society, electric power generation companies have adapted by building more steam-driven power plants. Lately, there has been a greater investment into solar and wind energy. But is that enough? And even if it was, is it the best and most efficient way to harness energy?

Due to a combination of political and technical influences, we are heading toward a crisis in energy delivery. Shifting political agendas desire the use of less oil, gas, and coal—the lifeblood of automobiles and power plants—and more use of public transportation and electric cars. Without turning to alternative means of generating electricity in the same quantity that fossil fuel-fired power plants produce, which in our world means *nuclear,* this would increase the demand for electricity while reducing the supply. Electric car manufacturers might serve their customers better and sell more cars if they can ensure a continued supply of electricity, using their genius to reinvent energy production and build their businesses around the results. Could an electric car eventually ditch its battery and be powered by a pyramid electron harvester, with relay stations positioned like cell phone towers?

Twenty-five years have passed since the publication of *The Giza Power Plant: Technologies of Ancient Egypt.* Since then, a lot of new information has been gathered by myself and other researchers in many fields, and I feel a need to report on how new discoveries fit with the power plant theory and play a part in bringing the vision described above into reality.

When *The Giza Power Plant* was first published, I didn't know what to expect. The book was akin to a reverse engineering exercise, and I was aware that other interpretations of the evidence used in my analysis may be used by others to arrive at different conclusions. I was also aware that there would be pushback based on historical grounds and long-held cultural ideology. I expected all of this and received it. After all, it was a theory—not presented as a fact.

Scientific theories must be falsifiable. They have to be tested and proven beyond a reasonable doubt that they are either correct or false. To do this requires the collection of evidence. In *The Giza Power Plant,* I presented an abundance of evidence that supported my theory that the

Great Pyramid was a power plant. I placed a stake in the sand of Egypt, so to speak, with the knowledge that any evidence discovered in the future would have the potential to either support or disprove the theory. Whatever is discovered cannot be ignored and must be addressed—even if it casts doubt on a particular theory that we may support.

A theory should also be able to accommodate new information that it did not anticipate, which, when added to the body of existing evidence, makes the theory more complete. In that sense the theory is predictive. And in that sense, *The Giza Power Plant* was predictive, and evidence that has been revealed since it was published supports it. Much of it is now well-known and discussed frequently on many social media outlets. Unlike the time when I was writing the book, between 1977 and 1998, in 2022 we can find that a tsunami of information about the Great Pyramid has made its home on the internet. For this reason, much of the research that appears in *The Giza Power Plant* is not going to be rehashed here. I don't want the reader to walk away saying this book is only offering the same thing, only in a different way. Moreover, I'm not going to make you wade through reams of technical reports in order to understand the overall picture. I will summarize main points, but will include appendices at the end written by experts where technophiles can dig through the details.

This book will introduce new evidence that may be considered predictive within the construct of the power plant theory, as well as new evidence that was not predicted, but which provides more support to the theory than what was previously known. The evidence is compelling, paradigm changing, and spectacular. It more than validates the power plant theory.

In this book you will learn:

1. How, since *The Giza Power Plant* was published, further explorations of the Queen's Chamber's southern and northern shafts have discovered evidence that supports their function described in the power plant theory. In particular, there is explosive new evidence that was discovered in 2002, but languished in a desk drawer until 2021, when it was broadcast on Matt Sibson's

Ancient Architects YouTube channel. What has been revealed fully supports the Giza Power Plant theory. Why this evidence has been hidden for so long is still an open question. Also, you will find an Appendix by Brett I. Cohen, PhD, who provides a different perspective and formula for the generation of hydrogen in the Queen's Chamber.

2. The discovery of scorch marks on the ceiling of the Grand Gallery, which supports the King's Chamber explosion hypothesis. While temples such as Denderah near Luxor had layers of soot that, when cleaned, revealed beautiful original paint and details, the ceiling of the Grand Gallery, when cleaned, revealed heat-affected areas burned into the limestone ratcheted ceiling.

3. The results of further acoustic testing and key measurements related to specific frequencies within the Great Pyramid's chambers and shafts. Robert Vawter's appendix will go into greater detail.

4. The stunning significance of the large void discovered above the Grand Gallery during muography detection efforts by an international group of scientific institutions. Aerospace Engineer Eric Wilson recognized immediately what this "Big Void" may have been created for and why it was discovered where is was. Also, his appendix is a paper he wrote and delivered to a conference of the AIAA.

5. The recent Russian research regarding how the Great Pyramid focuses electromagnetic energy to the center of the pyramid near the King's Chamber.

6. The exhaustive research into the phenomenon known as "earthquake lights" by a team of NASA scientists and the significance of what is known as the "Freund Effect" and how it relates to the function of the Giza Power Plant.

Of all those on the above list, the Freund Effect is probably the most significant of all, for I believe that it is these natural phenomena that inspired the building of the pyramids in the first place. Moreover, while the pyramids may have been stripped of the ability to continue to access and maximize the benefits of this effect, the effect is still there, dormant though it may mostly be, but what it may mean is that the Earth can be coaxed into providing our civilization with an abundant and inexhaustible source of electricity that will result in a more harmonious environment for future generations.

The results will affect more than one aspect of our lives as the physics at work in the Earth's crust and mantle have engendered a new understanding, and earthquake lights tell us not only where stresses are building before an earthquake happens, they also deliver another message—electrons are released when rock is stressed, and if we can build a system that simulates that stress and harvests the resulting electron flow, we can deliver it to the myriad devices we use on a daily basis.

Rock is the essential source of the electrons, and understanding how they become active under stress makes the mystery surrounding unexplained phenomena experienced around the pyramids more understandable. It explains much of the phenomena reported about electromagnetic activity in the Great Pyramid—such as the Leyden jar experiment, when Anglo-German engineer, metallurgist, and inventor, Sir William Siemens, after feeling a prickling sensation in his index finger when climbing the pyramid, crafted a Leyden jar and shocked his tour guide with a discharge of electricity the jar had accumulated.

The piezoelectric qualities of quartz-bearing granite being the source for electron flow in the King's Chamber will also be re-examined. Friedemann Freund's research into earthquake lights, and research in the field with experienced geologist, Adrian Lungan, who provided his own analysis and opinion, have resulted in more convincing and sustainable answers to questions that have arose.

It was 12 years after *The Giza Power Plant* was published before my second book, *Lost Technologies of Ancient Egypt*, was released. This book did not discuss the power plant theory but focused on another important aspect of ancient Egyptian technology: the means and methods with which they quarried and crafted their amazing stone artifacts.

Some have argued that this book logically should be read before *The Giza Power Plant*. In some ways, I would agree with them. Its contents are certainly less controversial and can be more vigorously defended. By the time it came to writing *Lost Technologies,* I had come to the realization that if humankind's understanding of the truth about the level of technology that was known and exercised in Egypt's ancient past was to change, then Egyptian history would have to be changed by the Egyptians themselves. To that end, I inscribed in *Lost Technologies of Ancient Egypt* a dedication to *"The Egyptians and their glorious heritage."*

An amazing transformation has taken place since. Although 25 years ago, when discussing my ideas on internet forums, I was told I was a lone voice in the wilderness with nobody listening or interested, there has been a grassroots movement of technologists, engineers, and scientists who have traveled to Egypt to examine the artifacts that I discuss in my books. The effect of this swell of interest and confirmation is that Egyptian tour guides who work with these knowledgeable and skilled travelers are now learning how advanced their ancestors were. My good friend and Egyptologist tour guide, Mohamed Ibrahim, has become a huge advocate for my work and is teaching his colleagues the truth about their ancestors using the instruments I left with him in 2019. I'm pleased to say that my books are becoming well-known in Egypt and receiving support not just from Egyptologist tour guides, but from engineers like Ahmed Adly who kindly provided the Foreword for this book, and mainstream scholars, such as Dr. Wassim El Sisi, a urologist and noted Egyptologist who discussed *Lost Technologies of Ancient Egypt* in a recent interview on Egyptian television and, recently, physicist A. L. Sherbini of Cairo University. Support is also provided by international law judge and legal overseer of the ScanPyramids mission, Hamada Anwar, as well as a leading geophysicist in Cairo, Taha Rabeh.

In 1986, during a visit to the Egyptian Museum in Cairo, my Egyptologist tour guide introduced me to a senior official. I gave him a copy of "Advanced Machining In Ancient Egypt?" This was an article I had published in *Analog Science Fiction and Fact* magazine in August 1984 (Dunn 1984). The official opened a desk drawer, tossed the article in, and wished me a good day. I never heard anything back from him. Today Egyptian minds are changing. I am no longer a "lone voice in the wilderness." Many others, both Egyptians and non-Egyptians, are inspired to lend their own voices to the subject.

My principal objective while visiting Egypt is not related to conventional Egyptology, archaeology, or rewriting Egyptian history, but to report on my own and others' observations. I began my research with a basic hypothesis that the Great Pyramid was a power plant and not a tomb. It was not until my first trip to Egypt in 1986 that I came face to face with the amazing manufacturing precision that is cut into their myriad of stone artifacts—whether these artifacts be intact statues or temples, or broken and discarded pieces surrounded by rubble. At that time, the purpose of my research expanded from a reverse engineering exercise to bringing Egyptians' attention to these special artifacts so that they may recognize the true genius of their ancestors. Should these further studies by qualified Egyptians result in a desire to rewrite their history, that should be up to them. Not outsiders.

The words I wrote in the final paragraph in the Summary chapter of *The Giza Power Plant* are as relevant today as they were more than two decades ago:

> We know very little about the pyramid builders and the period of time when they erected these giant monuments; yet it seems obvious that the entire civilization underwent a drastic change, one so great that the technology was destroyed with no hope of rebuilding. Hence a cloud of mystery has denied us a clear view of the nature of these people and their technological knowledge. Considering the theory presented in this book, I am compelled to envision a fantastic society that had developed a power system thousands of years ago that we can barely imagine today. This society takes shape as we ask the logical questions, 'How was the energy transmitted? How was it used?' These questions cannot be fully answered by examining the artifacts left behind. However, these artifacts can stimulate our imaginations further; then we are left to speculate on the causes for the demise of the great and intelligent civilization that built the Giza power plant (Dunn 1998, 225).

We are faced with a fascinating mystery. On the one hand, there have been observations in our atmosphere of a functioning machine that operates in a futuristically dazzling way. But we have no clue what technology is used that allows them to operate. Nor have we any clue how it looks on the inside. Then, surviving from deep in our past, we have the remains of a machine. We have not witnessed its operation. We know what it looks like on the inside, but it is so unique that there is nothing else like it on Earth. As much as we may want to, our technology is not sufficiently advanced to allow us to capture a UAP and study its assembly. We can, though, get closer to understanding how the Great Pyramid functioned, and I'm happy to say that I am much more confident today that I am on the right track than I was when *The Giza Power Plant* was first published.

What future will our children, grandchildren, and their grandchildren behold? Will the advances our generation have witnessed over the past 75 years continue into the future? If so, it would be impossible to imagine exactly how future generations will interact with their environment. My studies of the Great Pyramid over the past 44 years have given me a vision of what might exist, if the science and physics embodied in the Great Pyramid is recognized and taken seriously by those in authority who can motivate and inspire action to replicate the ancient Egyptians' accomplishments.

Looking into the past, as new information becomes available, we see a landscape that is littered with stories once believed and held dear now being abandoned for the sake of progress and truth. Stories that have been passed down for generations are reexamined and either retained, discarded, or redefined as youth asserts itself in the public debate—to the benefit of future generations. While honoring and respecting those who have come before, I dedicate this book to the new generation, those already born and those yet to be born, regardless of their location in the world. They will shape a different world and create a narrative with a vision that may be impossible for people today to even imagine. In *Giza: The Tesla Connection,* I am proposing that this is a world that was lived in before, and this is a world that can be lived in again.

ONE
THE GIZA POWER PLANT

PLATEAU FEVER

The mid-1990s was a very active time in the field of what was referred to at the time as "alternative history." It seemed that all eyes were focused on Egypt and the Giza Plateau. This was a time when John Anthony West, Robert Schoch, Graham Hancock, and Robert Bauval were extremely active on the plateau. Books published by these authors became popular and were accompanied by documentaries that were broadcast to large

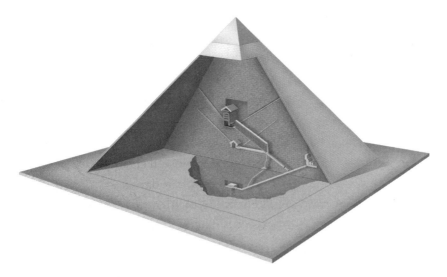

Figure 1.1. The Giza power plant.

audiences. The world, including numerous other researchers, gravitated to the Grand Central Station of ancient mysteries—the Giza Pyramids.

The chief nemesis of many of these alternative history writers was the charismatic and fiery Egyptian Egyptologist, Zahi Hawass, PhD, director of the Giza Plateau. Tension existed between Hawass and some Western writers and would-be writers. By the time I met Hawass, he had already been involved in controversies involving West, Hancock, Bauval, and others, and he even had other researchers accusing him of secret digging in the Great Pyramid, where his access was said to be gained through a secret tunnel that started in a bathroom in his office.

While much of this activity may have been irritating to Hawass, the effects of it were transformative for tourism to the area. The controversy was good for book sales and viewers of documentaries and website bulletin boards, and fans of numerous researchers flocked to Egypt to witness firsthand what they had been reading. Since I first visited Egypt in 1986, the facilities that cater to tourists have undergone a dramatic transformation. The airports have been radically revamped and made more efficient. Sites that were seldom visited, such as the Aswan Quarry where lies the Unfinished Obelisk, are now fitted out with visitors' centers, bookstores, gift shops, and cafes. When I first visited Aswan in 1986, none of that was there, and my fiancée and I were the only visitors to the site at that time.

There were few tourists to deal with in 1986. Entrance to the Giza Plateau had minimal security and the entrance cost was very low. There were very few tour buses compared to the crowds they bring today. Only the pyramids themselves have not changed. But what do we expect? They haven't changed in several millennia!

In September 2001, I was with Gail Fallen, field producer for Grizzly Adams Productions, when I first met Hawass. Our first meeting took place three years after *The Giza Power Plant* was published. Remembering the reception my article had received in 1986 at the Egyptian Museum, I expected a frosty response would be given to a book with a far more controversial message, so I did not present him with a copy. He was curious to know what I did and what I was going to speak about. I simply told him I was going to discuss the remarkable precision of the stonework found on the Giza Plateau and inside the pyramids.

Having heard the rumors that were swirling around the Giza Plateau at that time, I was very curious about the rumor that Zahi Hawass, while claiming he had closed the Great Pyramid for several months to perform maintenance and cleaning, was actually continuing to perform clandestine digging to access the end of the South Shaft of the Queen's Chamber. Following the filming of myself and Hawass (separately) on the Giza Plateau, Gail and I went to his office near the pyramids to finish up administrative work and take photographs. Hawass laughingly dismissed such rumors while generously giving us written permission to enter the then-closed Great Pyramid to see the work that had been done.

Figure 1.2. Zahi Hawass and Chris Dunn.

Figure 1.3. Zahi Hawass and Gail Fallen.

We also were given permission to see the Workers' Village, of which discovery Hawass was immensely proud, and special permission to visit the Serapeum. A full account of our visit to the Serapeum and the Workers' Village—along with commentary by Egyptian architect Hossam Aboulfotouh—is contained in *Lost Technologies of Ancient Egypt: Advanced Engineering in the Temples of the Pharaohs*. An update to the Serapeum section can be read in this book in chapter eight, "Saqqara's Hidden Treasures of Advanced Technology" where I discuss the question, "Where are the machines?"

A visit inside the Great Pyramid is inspiring at any time of day, whether morning, midday, or night. It is also impressive to behold whether it is covered in dirt and dust or not. On that day, however, I saw that the stated reason for its closure was true. Cleaning had taken place, and that cleaning had revealed, in the Grand Gallery, a tantalizing detail that caused me great excitement. It was a small but enormously important disclosure that has not been acknowledged by anyone else as being relevant but holds a prominent place in the list of evidentiary materials that underpin the power plant theory. This will be discussed in more detail in chapter four, "The King's Orchestra."

Regarding the proposed function of the pyramids as power plants, the science and technology built into their operation would most likely be considered proprietary if it were to be invented and implemented in today's corporate world. There would be no information available to the public about its interior design and various functions. In truth, there probably would not be much general public interest in venturing inside— although, depending upon the mindset of the ancient culture that supported their original construction, access to the pyramids' interiors may have been forbidden. More likely, it was impossible while they were operating. Because other pyramids of the same design, and presumed function, do not exist in other parts of the world, it is likely that the technologies that were developed and functioning in Egypt during that ancient period were guarded state secrets and not shared. But that is an opinion posited by a mind heavily influenced by modern Western values, so who knows?

It has long been held that Mystery Schools in ancient Egypt held knowledge that was only shared with certain initiates. Pythagoras, the Greek mathematician and philosopher, is said to have spent 22 years studying there, and upon returning to Greece, introduced to the West mathematical concepts that can be seen in ancient Egyptian structures today. While Pythagoras's name is now attached to the 3-4-5 right-angled triangle, the Pythagorean triangle is clearly evident in the proportions designed into ancient Egyptian statues. Other mathematical concepts that were introduced in the West by Greek mathematicians and philosophers are also evidenced in ancient Egyptian structures, such as the golden ratio (phi), pi, the ellipse/ellipsoid, and the Fibonacci sequence (Dunn 2010, 40–67).

REVERSE ENGINEERING—TRICKS AND SECRETS OF THE TRADE

The sharing of knowledge and enlightenment from people in one part of the world to people in another part of the world can be seen by an objective observer as a positive development. I know of no objections by the Egyptians to Pythagoras, a foreigner, exporting what he learned in the Mystery Schools in Egypt to his home country, Greece. Since the Industrial Revolution, though, the free flow of knowledge has become more restricted when affecting trade and a person's ability to make money. Trade secrets do not always end up being protected by a patent, and patents do not always completely halt the exportation of knowledge. A patent can provide inspiration for a competitor to improve on the usefulness of the invention and file for their own patent.

Exporting trade secrets from one group of people to another, whether permitted or not, has been going on since invention began. Perhaps the most famous case is the exportation of the cotton spinning mill from England to the United States by Samuel Slater in 1789. The machine brought by Slater to New York was not a physical object, nor was it a tube of rolled-up blueprints. It was all in Slater's head—though some have speculated that some secrets were hidden in his shoe. He had learned how the spinning mill was manufactured and operated while serving a seven-year apprenticeship under its genius inventor, Sir Richard Arkwright's partner, Jedediah Strutt.

Because it was illegal in England to export critical technologies, Slater disguised himself as a farm laborer and sailed from England while carrying nothing to outwardly indicate he was holding this protected information. Slater, celebrated as the "Father of the American Industrial Revolution," is considered a hero in the United States. No such accolades were awarded to him in his hometown of Belper, Derbyshire, where his angry neighbors referred to him as "Slater the Traitor." Time seems to have healed that wound, however, as the inhabitants of Belper subsequently forged a fruitful relationship with Pawtucket in New England, and a plaque honoring Slater now adorns his childhood home (White 1836).

Slater's exportation of technology might be considered a reverse or back engineering exercise. In the United States, he constructed what he had learned by closely studying how a machine, in this case a spinning mill, with its numerous parts designed and built by himself and others, functioned. The invention is credited to Arkwright, with some

Figure 1.4. Samuel Slater (June 9, 1768–April 21, 1835).

credit for his success given to previous inventors, but it is possible Slater, as Arkwright's apprentice during this period, may have contributed his own genius to its ultimate success and felt a degree of ownership.

Today, reverse engineering is a common process that is well-known around the world. Consumer product manufacturing companies understand that the beautiful, creative, and immensely functional new device they have put on the market will at some point end up in a competitor's engineering laboratory. A company may take a competitor's product—say, a vehicle—and observe its function, then take it apart, testing each individual part. Their ultimate success at creating a better version of their competitor's machine depends on their manufacturing capabilities and their ability to identify where they can make improvements.

Whether the originators of technology like it or not, the diffusion of manufacturing prowess around the planet has resulted in previously poor countries lifting themselves out of poverty and improving their lot in life. Perhaps this was Samuel Slater's motivation.

REVERSE ENGINEERING STEP ONE: EXAMINE, MEASURE, AND RECORD

Regarding the Great Pyramid, my theory that it originally served as a power plant resulted from a process of reverse engineering. It goes without saying that if we could take possession of one of these high-performance UAPs (aka UFOs), our engineers and scientists would attempt to reverse engineer it. Unlike the elusive vehicles we call UAPs, the Great Pyramid has been measured and examined closely inside and out by numerous explorers and researchers over centuries. While their understanding of the original purpose of the pyramids may have been in error, and their intentions may have been to find a hidden burial chamber, these researchers were conducting a reverse engineering task, whether they were aware of it or not. No other structure in the world has received anything like such attention over the centuries.

The list below represents the physical features that had been discovered and used when explaining the power plant theory in 1998. These details fall into three categories.

1. The discovered features were intentionally designed into the structure to perform a specific function.
2. The discovered features were a result of the machine's operation (similar to a buildup of carbon on an internal combustion engine cylinder head, for example.)
3. The discovered features were neither associated with the machine's original function nor were they the result of the machine's operation.

Each item on the following list will fall into one or more of the above categories, as indicated by [1], [2], or [3].

- Selection of granite as the building material for the King's Chamber. It is evident that in choosing granite the builders took upon themselves an extremely difficult task. [1]
- Presence of four seemingly superfluous chambers above the King's Chamber. [1]
- Characteristics of the giant granite monoliths that were used to separate these so-called construction chambers. [1]
- Presence of exuviae, or the cast-off shells of insects, that coated the chamber above the King's Chamber, turning those who entered black. [2 or 3]
- Violent disturbance in the King's Chamber that expanded its walls and cracked the beams in its ceiling but left the rest of the Great Pyramid seemingly undisturbed. [2]
- Theory that the so-called Guardians (generally identified as senior staff of Egyptian sites who held the keys to the temples) were able to detect the violent disturbance inside the King's Chamber, when there was little or no exterior evidence of it. [2 and/or 3]
- Reason the Guardians thought it necessary to smear the cracks in the ceiling of the King's Chamber with cement. [2 and/or 3]
- Fact that two shafts connect the King's Chamber to the outside. [1]
- Design logic for these two shafts—their function, dimensions, features, and so forth. [1]
- Reason for the Antechamber. [1]
- Grand Gallery, with its corbeled walls and steep incline. [1]
- Ascending Passage, with its enigmatic granite barriers. [1]
- Well Shaft down to the Subterranean Pit. [1]
- Salt encrustations on the walls of the Queen's Chamber. [2]
- Rough, unfinished floor inside the Queen's Chamber. [1 and 2]
- Corbeled niche cut into the east wall of the Queen's Chamber. [1]
- Two shafts that originally were not fully connected to the Queen's Chamber. [1]
- Copper fittings discovered by Rudolf Gantenbrink in 1993. [1 and 2]
- Green stone ball, grapnel hook, and cedar-like wood found in the Queen's Chamber shafts. [3]

- Plaster of Paris that oozed out of the joints inside the shafts. [1 and/or 2]
- Repugnant odor that assailed early explorers. [2]

The above physical features were discovered by early and more recent researchers. Other features of the Great Pyramid that influence the fundamental nature of the structure were related to its size, interior, and exterior dimensions and relationship to the Earth. Also contributing to the theory of the Great Pyramid being a power plant were research and observations related to the acoustics of the structure. There is, however, an aspect of all research that cannot be conveyed well in written or spoken words. This is the personal experience of the researcher and their tactile and mental interaction with the object they are studying. When it comes to a structure like the Great Pyramid, I cannot stress enough how important to me this aspect became over many years and multiple visits to Egypt.

I am not alone in articulating an almost magnetic attraction to the Great Pyramid. In their book, *Giza and the Pyramids: A Definitive History*, Zahi Hawass and Mark Lehner write:

Hawass believes there are chambers still hidden inside the pyramid, and these doors may be the keys that will open these secret rooms. Both of us have been inside the Great Pyramid many times. But it is when we are alone that the silence captures our hearts. Each chamber holds its own secrets, and each time you enter, you feel you are part of an adventure in history. When the robot went inside the tunnels in the Queen's Chamber, revealing how the stones of the interior are interlocked, it reminded us of what the Arabs said about the pyramids: 'Man fears time, but time fears the pyramids.' (Lehner and Hawass 2017, 163)

The connection with a particular place must have been the fundamental reason why the Great Pyramid and its neighboring pyramids were built where they were. One of the most recognized facts about the builders of the pyramids is that they were, at a minimum, planetary earth scientists. Today we can draw that conclusion when we examine the pyramid's relationship to the earth. Its effect on visitors within its chambers may seem like a mysterious or magical experience, but as will be revealed in a later chapter, Arthur C. Clarke's maxim may be at work here also. There may be a legitimate underlying physical aspect of nature that can be described in an evidence-rich way that may satisfy the strictest adherent to the scientific method.

While our list of evidence includes the discoveries of early explorers who revealed physical aspects of the Great Pyramid, another breed of researchers applied themselves to collecting measurements and investigating what those measurements represented. Help in looking for answers came from those who may or may not have lent a hand in collecting the information, but studied what was collected and added their own analysis. In 1859, the publisher and author John Taylor (1781–1864) published *The Great Pyramid: Why Was It Built? And Who Built It?* In performing his analysis, he used measurements taken by an English mathematician with a passion for measuring ancient monuments, John Greaves (1602–1652), as well as those taken by the French savants who studied the Great Pyramid during Napoleon's expedition to Egypt (1798–1799).

Taylor's conclusions were revolutionary for his time, but based on the measurements available to him, he stressed that the intentions of the builders of the Great Pyramid were to reflect a knowledge of the geometry and measurements of the Earth. "What reason, it may be asked, can be assigned for the founders of the Great Pyramid giving it this precise angle [of 51° 50'], and not rather making each face an equilateral triangle? The only one I can suggest is, that they knew the Earth was a sphere; and by observing the motion of the heavenly bodies over the Earth's surface, had ascertained its circumference, and were desirous of leaving behind them a record of the circumference as correct and imperishable as it was possible for them to construct" (Taylor 1859, 19).

Though not having visited the Great Pyramid himself, Taylor intuitively recognized that the work performed by John Greaves was not just a whimsical fancy that filled his time. Taylor recognized Greaves's mission: to reveal the truth. Taylor writes:

The wise and careful observer, from whose work we have made these large extracts, had a feeling of the future value of his calculations, which seems to have been almost prophetic. He does not tell us what expectations he entertained; but since they led him to take every precaution to ensure the most minute accuracy in all his measurements, he must have made them with a view to some such issue as that on which we are now bringing them to bear. He has left nothing unnoticed; he has omitted no remark which could in any way tend to elucidate the purpose for which the pyramid was built. As if he had supposed it to have been intended to serve as a standard of measure, he has anticipated every question which could have been put by any one [sic] whose object was to prove the case which we are endeavouring [sic] to support . . . . (Taylor 1859, 114)

Taylor's work aroused the interest of Sir William Flinders Petrie, who performed more-accurate measurements a decade later. That Petrie's measurements differed from Greaves's is not surprising considering that Petrie, the son of an engineer, was a trained surveyor and as such was more qualified than Greaves to take measurements using state-of-the-art metrology equipment of that period.

From experience, I can say a decade makes a lot of difference in what becomes available in terms of not just knowledge but also the introduction of new tools to assist in gaining more accurate knowledge. In 2001, the instruments I took into the Serapeum were simple rudimentary gauges: a precision toolmaker's square and a straightedge. Over a decade later, Egyptian engineer Ahmed Adly procured a Bosch digital laser distance measure and a digital protractor. I have been retired for 10 years, but in 2023, my son, Alex, if he gained access and permission to do further studies of these boxes, might use a Hexagon AS1 scanner with a Leica AT960 laser tracker to gather a three-dimensional point cloud that could be downloaded by anyone to study. Time and technology move on, and who knows what will be available in 2033?

It is important to remember, though, that while today's measuring instruments are the results of decades-long development and are superior to the instruments Greaves and Petrie used to take their measurements, theirs were superior to those found in the archaeological record and displayed in museums.

Was the ancient pyramid builders' knowledge of our planet limited to just its physical dimensions or did it include the nature, concentration, and movement of the energy it contained? Assuming the answer to that question was yes and their knowledge of the Earth did not end at its surface, I wrote in *The Giza Power Plant*:

> As electrical energy can create mechanical vibrations (perceived as sound by the human ear), so in turn can mechanical vibrations create electrical energy, such as the previously mentioned ball lightning. It could be theorized, therefore, that with the Earth being a source for mechanical vibration, or sound, and the vibrations being of a usable amplitude and frequency, then the Earth's vibrations could be a source of energy that we could tap into. Moreover, if we were to discover that a structure with a certain shape, such as a pyramid, was able to effectively act as a resonator for the vibrations coming from within the Earth, then we would have a reliable and inexpensive source of energy. (Dunn 1998, 130)

PLEASE, EARTH, CAN YOU SPARE US A FEW ELECTRONS?

As it turns out, when I wrote *The Giza Power Plant*, my perception of the full nature of the energies that influence the Giza Pyramids was insufficient. There is now much more information available to add to the body of knowledge supporting the power plant theory. That new information now provides me with much greater confidence than before that an untapped source of green and clean energy exists under our feet. Perhaps we could reduce our reliance on fossil fuels and eliminate many expensive and destructive practices just by coaxing the Earth to give up not oil or coal or gas but only the electrons we need.

Create a pact with the Earth. We will stop burning you to create steam to rotate turbine blades that rotate generators from which we harvest electrons, if you will provide us with just the electrons. Considering the number of electrons involved in the generation of steam, how could the Earth pass up such a deal?

In the 1970s, we were told that in about 20 years we would be receiving electricity from nuclear fusion power plants. Then we breezed through the 1990s, the 2000s, and the 2010s, and here we are in the 2020s still waiting for cheap unlimited amounts of fusion energy. But not to worry, they are working on it, and according to an article in *Forbes* magazine published in 2019, we should have it by around 2050:

> The International Thermonuclear Experimental Reactor (ITER) project under construction in Cadarache, France, is the most celebrated Tokamak-style reactor in existence. The multi-billion dollar, 35-nation effort including the United States, Russia, China, India, the European Union, Japan and South Korea is now on pace for a 2050 commercial debut after a number of cost overruns and delays. (Cohen 2019)

Nuclear fusion is the most desirable type of nuclear energy. Sadly, though, we don't have any. The fission plants in use today are viewed with suspicion by environmentalists and others, who point to the dangers involved when things go wrong. Certain examples provide a basis for concern, such as the infamous Chernobyl accident, which laid to waste a region north of Kiev in Ukraine. Then there was the dramatic evacuation of residents when radiation escaped the Three Mile Island nuclear power plant in the US after the Unit 2 reactor suffered a partial meltdown in 1979. The most recent and dramatic example that continues to provide fuel to the anti-nuclear-fission environmentalists' fire is the Fukushima Daiichi nuclear disaster—a perfect example of where not to build a nuclear power plant. Earthquakes, tsunamis, and nuclear power plants do not mix!

It seems we are in a bit of a predicament. Will we eventually be forced to stop burning fossil fuels? There are those of the opinion that we have barely scratched the surface of the Earth to exhaust it of its resources. Nonetheless, others wonder whether we should do it gradually and willingly or wait until there is nothing left? Some reports claim that in 2016 the world consumption of coal was estimated to be 8,561,852,178 short tons. With estimated world reserves at 1,139,471,430,000 tons, this means that in 133 years, if new sources are not identified, the world will run out of coal (Worldometer). Mining, transporting, and processing coal relies on numerous ancillary industries and incurs enormous cost. The future of oil is even more desperate, with some sources estimating that, with known reserves currently being 1,650,585,140,000 barrels and consumption averaging 35,422,913,090 barrels per year, the world will run out of oil in 47 years unless significant new sources of oil are found. (Worldometer). Fossil fuels are finite resources and eventually may not be available to us in the quantities used today or projected to be used in the future. How accurate these figures are remains to be seen, but even an optimistic vision of the future leaves you with the sense that the clock is ticking. What are we doing about it?

Reducing the demand for oil by using more electric vehicles does not solve the problem if, to generate more electricity, we have to burn more coal and oil. Oil that is refined to create gasoline that is no longer needed to power vehicles could be redirected to power plants, but how many more years will that provide to us?

For over 100 years we have primarily been living in a fossil-fuel-fired Steam Age. In the 1950s, following advances in nuclear physics, power companies added nuclear fission to their arsenal of electricity-producing fuels. However, despite the source of heat, the process of creating electricity was the same: Boil water to create steam, which, under pressure, pushes and rotates turbine blades. Our major electricity-generating technologies are based on the discovery of fire by primitive early humans. Even modern jet engines used to propel millions of people all over the world every year are simply a sophisticated use of fossil-fuel-consuming fire.

When I wrote *The Giza Power Plant,* I tried to gauge the response of readers. One of the options I considered was to write the book as a science fiction novel. This was because the conclusions I was coming to, when compared with the current state of accepted answers regarding our past, and particularly about the Great Pyramid, really did sound like science fiction or, as Professor Rubinstein put it, "straight out of Fruitcake Land"—and to many, it still does.

Ultimately, I decided not to shade my convictions, and I wrote: "Let me make no apology for the theory I am proposing. The Great Pyramid was a geomechanical power plant that responded sympathetically with the Earth's vibrations and converted that energy into electricity" (Dunn 1998, 151).

HARVESTING A BUMPER CROP OF ELECTRONS

Today, with the valuable list of the evidence that has been revealed since, including the recently discovered "Big Void" and features revealed in the Queen's Chamber's North Shaft, my previously strong convictions are even stronger. While making a slight change to the wording, I'll say again that I make no apology for the two questions I am asking:

1. Can an abundance of electrons be harvested without burning fossil fuels or creating nuclear reactions?
2. Were the ancient pyramid builders simply farming electrons with their pyramids? Could the pyramids be described as electron harvesters?

Electron harvesting is not a new concept. A quick internet search provides many sites that discuss the topic. I presented the above as two questions, rather than one, because even if we are not yet ready to accept the premise surrounding question two, I believe that to support our desire to reduce our use of fossil fuels and develop new sources of energy, question one should be given serious consideration.

A proposal that pyramids are a technical solution for our constant need for electricity introduces a level of complexity when considering the pyramids' place in history books. While new sources of energy, along with technologies for their extraction, would bring the world relief, there is strong resistance to rewriting our history books if Egypt's pyramids are recognized for what they truly are.

Nonetheless, according to Ahmed Adly, a quiet revolution of ideas in Egypt is already affecting the Egyptological community. This is a very positive development, for sustaining a mainstream belief system that is lacking in evidence becomes harder and harder as the population becomes more educated. A growing number of professional young people in Egypt are now questioning the stories they have been told about their history. Some, like Ahmed Adly, Mohamed Ibrahim, Yousef Awyan, and others are employing hard scientific principles and modern instruments to reexamine the evidence.

Egypt's youth are taking the reins and leading the charge to restore the truth about their ancestors. They are pointing out that what they were taught in school were stories regarding Egypt's past that were formulated in the Western world by scholars, astronomers, engineers, and philosophers from the Victorian era to the present, and they are passionate about correctly interpreting their ancestors' brilliant accomplishments and elevating the prestige of Egypt in the annals of history. At the same time, they are doing it respectfully and kindly, without animus. In the next chapter, we will examine new evidence that sheds light on the path toward realizing Tesla's dream of wireless "free" energy while at the same time correcting a misunderstanding of world history.

TWO
THE GIZA-TESLA CONNECTION

We are whirling through endless space, with an inconceivable speed, all around everything is spinning, everything is moving, everywhere there is energy. There must be some way of availing ourselves of this energy more directly. Then, with the light obtained from the medium, with the power derived from it, with every form of energy obtained without effort, from the store forever inexhaustible, humanity will advance with giant strides. The mere contemplation of these magnificent possibilities expands our minds, strengthens our hopes and fills our hearts with supreme delight.

NIKOLA TESLA

Because my research in the past focused principally on the Great Pyramid and paid little attention to all other pyramids in Egypt, I have been accused of suffering from what some critics have called "Great Pyramid isolation syndrome." This is evidently a mental condition experienced by those who focus on just the Great Pyramid to the exclusion of over a hundred other pyramids in Egypt. My simple response to those accusations is: With no, or slim, evidence, Egyptologists claim that all of Egypt's pyramids were built to be tombs. With an abundance

Figure 2.1. Tesla's Wardenclyffe Tower.

of robust evidence, I will theorize that all of Egypt's pyramids were built to serve a more practical purpose, that is to harvest electrons contained in igneous rock in the Earth's lithosphere.

The fundamental science that supports the Giza power plant theory is the same understanding of nature that is embodied in every other pyramid in Egypt. It is, as Ahmed Adly specifically noted, *"the power generated from an exceptional understanding of the physical properties of the rocks."* While Adly was referring to the ancient Egyptians' presumed knowledge and understanding of rocks, today, fortunately, we have scientists who, in my opinion, have recovered that lost knowledge—and what they have discovered has the potential to transform the planet in ways we cannot even imagine.

WHEN ROCKS GO CRUNCH

When it comes to landscaping, the immense forces that create breath-taking natural wonders make our endeavors seem insignificant and small. Reminders of these awesome and destructive events occasionally shake us from our peaceful enjoyment. One such reminder came on December 26, 2004, at around 7:58 a.m. local time (UTC+7), when innocence on the shores of South East Asia was stripped away as a large region of the ocean floor where tectonic plates meet suddenly thrust upward, lifted the water above, and caused a devastating tsunami. Vacationers near the water's edge in a bay area were puzzled when the sea receded and revealed acres of sand; some ventured out toward the ocean before the sea returned with a vengeance. Wave heights up to 100 feet (30 meters) engulfed these and other happy vacationers who were simply enjoying the idyllic beauty of a paradise provided by nature and modern humans. Fourteen countries were tragically affected by the tsunami, during which an estimated 228,000 people lost their lives (Pacific Coastal and Marine Science Center 2018).

The Sumatra-Andaman earthquake, as it is called in the scientific community, occurred in the Indian Ocean north of Simeulue Island in an undersea fault line where the Burma and India plates converge. Near the western coast of northern Sumatra in Indonesia, a vast amount of energy measuring 9.1 on the Richter scale and equivalent to 480 megatons of TNT (32,000 Hiroshima-size bombs) was released. If harnessed, what was unleashed when the rocks went crunch contained enough energy to power every five-star hotel overlooking every beauty spot in the world for many years to come.

Preceding it three days earlier, a different type of earthquake shook buildings on the island of Tasmania off the south coast of Australia. That earthquake, centered 305 miles north of Macquarie Island, was the largest recorded since an 8.4 tremor shook the coast of Peru on June 23, 2001, killing 74 people. Had this earlier quake been the result of vertical displacement, as in the Sumatra-Andaman quake, then most assuredly the shores of southwest Australia and northeast Antarctica would have been awash with disaster.

A significant earthquake happened in the region of Abruzzo, central Italy, on April 6, 2009, at around 1:32 a.m. (3:32 a.m. local time). The L'Aquila earthquake is noteworthy for several reasons.

- Public safety workers, officials, and scientists who were responsible for alerting the citizenry to the impending destruction did not issue any alerts, and 308 people lost their lives.
- Seven members of the Italian National Commission for the Forecast and Prevention of Major Risks, six scientists, and one ex-government official were accused of giving inexact, incomplete, and contradictory information about the danger of the tremors that preceded the main quake. On October 22, 2012, each was sentenced to six years in prison, a sentence that was reversed two years later. Poor building standards were also given as a reason for the tragic loss of life.
- Precursors to the quake were detected by NASA satellites collecting infrared data. Several NASA scientists working out of the Ames Research Center in California had been studying electromagnetic precursors and a natural phenomenon called earthquake lights for over 10 years before the occurrence of the L'Aquila quake.

In *The Giza Power Plant*, I described the King's Chamber as a resonant chamber that, because of its material makeup, converted mechanical energy (vibration) to electrical energy. I quoted the work of other authors who claimed that the Aswan granite out of which the chamber was constructed was composed of feldspar and up to 55% silicon quartz crystal. With the quartz crystal subject to vibration, I theorized that the piezoelectric effect of the crystal was responsible for generating electron flow.

Because of the features and dimensions of the North Shaft that make it suitable for use as a waveguide for microwave energy, I posited that the King's Chamber complex and its shafts connecting to the outside would serve as a maser (microwave amplification through stimulated emission radiation.) The key evidences to support this theory are the North Shaft, with dimensions suitable for hydrogen radiation, and a bulbous opening in the south wall, shaped like a microwave horn antenna, that collected the amplified energy stimulated in the resonant King's Chamber.

However, with the revelations of Freund's new physics, to only identify the piezoelectric properties of quartz to explain electron flow from granite is limited in scope. In order to activate piezoelectricity, quartz has to be stressed across a specific axis, and the random orientation of crystals in the granite do not support the idea that piezoelectricity was a significant factor. While questions persist regarding the percentage and orientation of quartz in granite, with estimates ranging from less than 20% to over 60%, the entire discussion may be irrelevant.

In March 2018, I cohosted a Lost Technologies of Ancient Egypt tour with sound engineer Robert Vawter and Egyptologist tour guide Mohamed Ibrahim. A think tank of guests was attracted to the tour: aerospace engineer Eric Wilson, entrepreneur Carmen Miller, and numerous other highly accomplished professionals, one of whom was geologist Adrian Lungan, who graduated with first-class honors from the University of Western Australia and has had a successful 40-year career in the mining industry.

Adrian quickly impressed the guests with his quiet intelligence and his broad and specific knowledge of rocks. I was particularly impressed with his study methods at the Aswan quarries, when he picked up a piece of granite, studied it closely under a lens, and then put it to his mouth and licked it. Employing his taste along with sight and touch provided him with more information about the material makeup of the rock. Upon reflection, I was curious about this method, so I emailed him and asked about this as well as his analysis of the amount of silicon quartz crystal in Aswan granite. On August 31, 2021, he responded as follows:

> The Aswan granite is an acid-syenite which has a very specific composition in that the quartz content is very low (<5%). Robert Vawter and I discussed the piezo-electric properties of the quartz

in the Aswan granite and I pointed out that just because a rock contains quartz, it doesn't mean that all quartz in say, a granite, exhibits piezo-electric properties (i.e. mono-crystalline vs amorphous quartz) and with such a minor amount of quartz in the granite, I don't believe that the electrical properties would in any way be highly significant. I have attached a paper that covers something on this topic.

On licking rocks, when a rock is altered e.g. via hydrothermal fluids or by weathering, then certain clays can form. By seeing if your tongue sticks to a clay-rich rock, one can determine if the clay is kaolin (i.e. the rock has been kaolinised) or not and one may be able to work out what the clay was probably derived from and hence, the precursor. Also, when one crunches, say, a siltstone between the teeth, it feels 'gritty' when moving your teeth from side to side, whereas a mudstone feels smooth so, when not apparent, you can bite off a small fragment of the rock and use your teeth to determine the difference between the two rock types. Smell, commonly used for sulphides (if not already visible).

Some rocks you definitely don't lick, e.g. rocks with brightly colored minerals, e.g. mercuric minerals such as cinnabar or arsenical minerals. Many of these poisonous minerals, except for scorodite (As), are brightly colored, reds, oranges and yellows, so when you see distinct minerals with such colors, it's not a good idea to lick the rock. Most geologists should recognize them. (Adrian Lungan email received on August 23, 2021)

ONE DOOR CLOSES AND ANOTHER OPENS

While the possibility of piezoelectricity from granite became more remote, I was convinced that something else had created the electric phenomena experienced by many visitors to the Great Pyramid. Sir William Siemens, an Anglo-German engineer, metallurgist, and inventor, experienced a strange energy phenomenon at the Great Pyramid when an Arab guide called his attention to the fact that, while standing on the summit of the pyramid with hands outstretched, he could hear a sharp ringing noise. Raising his index finger, Siemens felt a prickling sensation. Later on, while drinking out of a wine bottle he had brought along, he experienced a slight electric shock. Feeling that some further observations were in order, Siemens then wrapped a moistened newspaper around the bottle, converting it into a Leyden jar. After he held it above his head for a while, this improvised Leyden jar became charged with electricity to such an extent that sparks began to fly. Reportedly, Siemens's Arab guides were not too happy with their tourist's experiment and accused him of practicing witchcraft. Peter Tompkins wrote, *"One of the guides tried to seize Siemens' companion, but Siemens lowered the bottle towards him and gave the Arab such a jolt that he was knocked senseless to the ground. Recovering, the guide scrambled to his feet and took off down the Pyramid, crying loudly"* (Tompkins 1971, 278–279).

Magic seems to be a popular common cause for things remarkable but misunderstood. Consider again Arthur C. Clarke's quote that "A sufficiently advanced technology is indistinguishable from magic." Rather than being labeled as something supernatural, these phenomena might be best explained by gathering geophysical/geoelectrical data from the Giza Plateau and comparing it to the results of research conducted by a physicist, Friedemann Freund, PhD, at NASA's Ames Research Center located in Silicon Valley, 12 miles north of San Jose, California. Answers to the mysteries of the pyramids may not all be found in Egypt. The research results of Freund and his associates may add valuable understanding about this phenomenon that influenced travelers to the Great Pyramid. Robert Vawter, now understanding after his return from our Lost Technologies of Ancient Egypt tour in 2018 that the piezoelectric effect does not play a significant role in the functioning of the Giza power plant, contacted Freund after accessing some of his research papers on earthquake lights on the internet.

Freund developed his theory during 24 years of research into the physics behind what have historically been known as earthquake lights. These lights appear in areas where there is an increase in tectonic stress, and in many cases, the lights will precede earthquakes. In the past, they have manifested as atmospheric electrical phenomena, such as ball lightning—which has sometimes been mischaracterized as UFO activity. Through satellite data analysis and laboratory experiments, Freund has discovered the underlying physics that explain how specific types of rock formations produce electrical anomalies in the Earth before and during earthquakes. Previously described in geophysics as earthquake lights, this phenomenon is now known by those associated with Freund as the Freund Effect. Freund's fundamental discovery was that igneous rocks in the Earth's mantle and crust contain dormant electronic charge carriers, and when stressed, certain types of igneous rocks essentially turn into a battery. See color plates 1–5 and Appendix B for details on his theory and discoveries.

Freund's motivation was not to solve the riddle of the pyramids, which was probably the furthest thing from his mind, but to examine a natural phenomenon that occurs prior to a seismic event, with the hope of being able to forecast when an earthquake is about to occur in order to provide adequate warning that could save lives. The evidence he has gathered to support his conclusions is impressive. The earthquake at L'Aquila is a prime example of where this type of early warning detection could be applied. NASA scientists, monitoring satellite infrared detection data, observed strong infrared emissions in the hills far from the earthquake epicenter around the region of Abruzzo, Italy, several days prior to what is now known as the L'Aquila earthquake. This energetic discharge is directly related to the phenomenon known as earthquake lights. The infrared emissions and the earthquake lights, according to Freund, are created by the release of energy via positive electron holes from peroxy bond defects when rock is put under pressure.

These aspects of the Freund Effect are technical geophysical terms that may seem opaque to the layman, but what they mean is that under stress, underground igneous rocks of different types can release massive amounts of electrons, which quickly migrate to high points on the surface where they ionize the air, creating earthquake lights. The energy necessary to achieve this is on the order of tens of kilovolts per square inch. Considering this discovery through the lens of the Giza power theory, the Freund Effect seems to identify a source where an inexhaustible supply of electricity exists and, under the right conditions, can be stimulated to move and be harvested for use for the same reasons we now mine, burn, and blow.

Using samples of different kinds of igneous rock, such as granite, basalt, and gabbro, Freund has demonstrated these phenomena in an Ames Research Center laboratory. Of note, his results identified gabbro as responding with greater efficiency at releasing electrons than granite. This is an important discovery, as it essentially dispels the idea that piezoelectricity plays a major part in creating earthquake lights. Gabbro does not contain quartz crystals.

Lightning is an example of electrons released into the atmosphere. What if we learned how to capture those electrons and use them to power our devices? What if we mimicked the random, unpredictable, natural forces of the Earth that stimulate the release of the electrons in a way that is predictable and enables a constant, generous supply of electricity?

Freund's theory regarding the physics behind earthquake lights is the result of years of research spent collecting and testing different types of rock in his laboratory at NASA. He and his associates also learned that the rock only has to move a few angstroms (1 angstrom = 10^{-10} m or 0.00000001 cm or 0.000000003937008 in.) for this phenomenon to occur. To bring this into clearer perspective, a human hair is approximately 0.0025 inches (0.0635 mm) thick. Divide a strand of hair into 635,000 parts and you will have one angstrom. The Earth is a dynamic body that is constantly in motion. It is constantly releasing electrons into the atmosphere, and according to Freund's research, it does not require much movement of a suitable type of rock in order to do so (Freund 2003, 2013).

Appendix B in this book is a transcript of Freund's lecture in Christchurch, New Zealand, in 1996, in which he explains his theory in layman's terms and asks the question, *"What if we could know when an earthquake was going to strike, days or even weeks in advance?"* The foundation for the theory is semiconductor physics and the existence of peroxy defects in igneous and metamorphic rock. He describes the phenomena as *"A new chapter in earthquake physics. A new chapter in materials science and solid-state"* (Freund 2016).

CONFIRMATION OF FREUND'S NEW PHYSICS

Although Freund's motivation behind his research into the physics under-lying earthquake lights was how that knowledge may be used to predict when an earthquake was going to occur, seismologists have not univer-sally recognized its value to them. Freund and his associates at NASA, along with the GeoCosmo Research Centre, were lone voices in geophys-ics and seismology until confirmation of their discovery was published by Japanese researchers who had replicated their work, which included, of course, their own laboratory experiments (Takeuchi et al. 2011).

WHEN ROCKS GO HUM

From my perspective, this discovery is enormously important to our understanding of how the Giza power plant operated. Combined with the acoustic design of the interior chambers and passageways of the Great Pyramid, everything points to the granite-lined so-called King's Chamber as being the power center where acoustic and electromagnetic energy pumped hydrogen atoms to a higher energy state. Freund's research fills in the void left by the rejection of the idea that quartz-bearing granite, when caused to oscillate, could produce sufficient elec-tron flow through the piezoelectric effect. With supporting evidence produced through laboratory experiment, his research appears to pro-vide an alternate mechanism based on semiconductor physics for har-vesting electricity and answers the question of why it was necessary to quarry and transport thousands of tons of granite 500 miles and pre-cisely shape it before assembling it in the center of the Great Pyramid.

In 1752, Benjamin Franklin, probably the modern era's first famous electron harvester, tapped into lightning using a key attached to a kite that was blown aloft during a thunderstorm. When threads on the wet hemp kite string were standing out straight with negative charge, Franklin moved his knuckle near the key and was rewarded with an electric spark and a shock (The Franklin Institute). Testing a theory and proving it correct while being a component in the materials used to test it was no doubt a bittersweet experience for him.

In 1879, Thomas Edison created the first dependable electric light bulb, and in the 1990s Freund started on his quest to understand the physics behind earthquake lights. Note that just over 100 years separates each endeavor. During this time, measurements, studies, and important discoveries were being made inside and outside the Great Pyramid in Egypt.

The industrial landscape has changed significantly in the past 100 years. If we project our minds 100 years into the future, what invention will be seen as having the most effect on society? Solid-state electron harvesting megamachines, perhaps?

The Giza Plateau is what fellow researcher Robert Vawter describes as a "strong area of interest" for what has become known within the GeoCosmo group as the Freund Effect. As certain types of rocks are compressed (put under a load or stressed), electrons flow out of the rock and travel quickly to the topographical high ground of the terrain. Electrons moving deep beneath the Earth can be redirected upward to the surface if they encounter a basalt dyke (a near-vertical crack, fault, or joint in the bedrock that allows molten magma to push to the surface), which acts as a wave guide.

The Giza Plateau has several of these dykes, and a future geological survey is planned on the plateau to measure the electromagnetic environment in the area. The work will be conducted by the National Research Institute of Astronomy and Geophysics near Cairo under the directorship of Professor Taha Rabeh. Knowing the natural energies and frequencies of the bedrock in that location could provide us with a greater understanding of phenomena that travelers have experienced when visiting the Great Pyramid—Siemens's Leyden jar experiment being one of them.

WHEN THE FREUND EFFECT IS FACED WITH A MOUNTAIN OF PRECISELY CUT STONE

Could the Freund Effect inspire the development of technologies that have the potential to provide us with the solution to our future energy needs? Well, it is inspiring me to discuss it, so it will be interesting to see how far it will go. It should be exponentially more attractive to power companies if they begin to understand the potential savings they would gain over current methods. The Giza Plateau's topography would seem to be the perfect candidate for maximizing the Freund Effect, though probably not quite as suitable before the pyramids were constructed as after. Now, we see artificial mountains standing with an open invitation for electrons released from deep in the Earth to rush to their peaks.

The Pyramids Exert Their Own Influence on the Electrons

In 2018, I started to receive a lot of emails regarding a recent scientific study, which concluded that the Great Pyramid, by its shape and construction materials, will focus electromagnetic energy to its center. In general, the emails I received related this research to my book, *The Giza Power Plant*, and regarded the discovery as supportive of the power plant theory. In summarizing their paper, "Electromagnetic Properties of the Great Pyramid," the authors write:

> Resonant response of the Great Pyramid interacting with external electromagnetic waves of the radio frequency range (the wavelength range is 200–600 m) is theoretically investigated. With the help of numerical simulations and multipole decomposition, it is found that spectra of the extinction and scattering cross sections include resonant features associated with excitation of the Pyramid's electromagnetic dipole and quadrupole moments. Electromagnetic field distributions inside the Pyramid at the resonant conditions are demonstrated and discussed for two cases, when the Pyramid is located in a homogeneous space or on a substrate. It is revealed that the Pyramid's chambers can collect and concentrate electromagnetic energy for . . . both surrounding conditions. In the case of the Pyramid on the substrate, at the shorter wavelengths, the electromagnetic energy accumulates in the chambers providing local spectral maxima for electric and magnetic fields. It is shown that basically the Pyramid scatters the electromagnetic waves and focuses them into the substrate region. The spectral dependence of the focusing effect is discussed (Balezin et al. 2018).

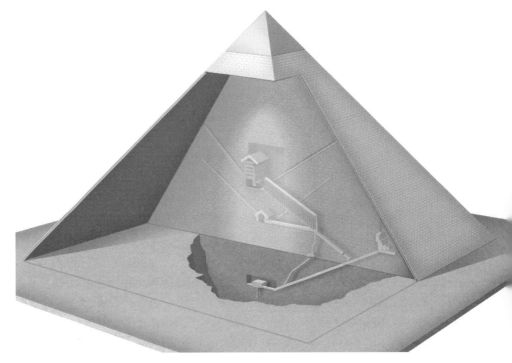

Figure 2.2. The Great Pyramid showing a concentration of electromagnetic activity at the power center. (See also color plate 6.)

The timing of these important pieces of research came to my attention during what I would consider an extremely short time span. This is quite stunning when you consider how slowly research information about the Great Pyramid has been forthcoming over the decades.

Anecdotal testimony of an electrical phenomenon associated with the Great Pyramid has been discussed for many years.

- Friedemann Freund's research identifies the source where an abundance of electrons can be coaxed from the Earth and the underlying physics that is responsible.
- A team in Japan confirmed Freund's discovery.
- A team in Russia provided the results of a study that shows how we can expect these electrons to behave within The Great Pyramid.

These separate research efforts have transpired independent of one another, and they provide a scientifically solid picture of a phenomenon that the physical evidence suggests took place. While there is still significant research to be accomplished, the scientific underpinnings now exist to explain the phenomenon known as earthquake lights. This, combined with other discoveries, fills gaps in the Giza power plant theory and brings it into greater focus and clarity. Are there other voids that still need to be explained? As it happens, there are one or two.

SCANPYRAMIDS MISSION DISCOVERY INSIDE THE GREAT PYRAMID

On November 2, 2017, a peer-reviewed article appeared in *Nature* magazine regarding a unique study of the Great Pyramid by an international team of scientists who set up muon detectors in the Queen's Chamber. Muon particles shower the Earth after being created through the interaction of cosmic rays with atoms in Earth's upper atmosphere. In the article, titled "Discovery of a Big Void in Khufu's Pyramid by Observation of Cosmic-Ray Muons," the authors describe an intriguing discovery:

> Here we report the discovery of a large void (with a cross section similar to the Grand Gallery and a length of 98.4 feet (30 m) minimum) above the Grand Gallery, which constitutes the first major inner structure found in the Great Pyramid since the 19th century. This void, named ScanPyramids' Big Void, was first observed with nuclear emulsion films installed in the Queen's chamber (Nagoya University), then confirmed with scintillator hodoscopes set up in the same chamber (KEK) and reconfirmed with gas detectors outside of the pyramid (CEA). This large void has therefore been detected with high confidence by three different muon detection technologies and three independent analyses. These results constitute a breakthrough for the understanding of Khufu's Pyramid and its internal structure. While there is currently no information about the role of this void, these findings show how modern particle

physics can shed new light on the world's archaeological heritage. (Morishma et al 2017, 386–390)

Curiously, despite the rigorous academic and scientific process behind the ScanPyramids mission's release of information regarding what their muon detection plates had revealed about the hidden interior of the Great Pyramid, some Egyptologists denounced the find and claimed that what was discovered was already known to exist and was not new. They complained that they were not consulted about the discovery or allowed input into what was revealed to the public (Moore 2017).

Later I will discuss the exploration of the Queen's Chamber South Shaft, when a breadbox-sized space was discovered, to much fanfare, behind Gantenbrink's door and announced to the world as an amazing discovery. I find it interesting that this small space somehow deserves greater recognition than a "newly discovered" space almost the size of the cabin in a Boeing 707 jetliner.

Notwithstanding the politics at play during this exploration, the science is solid, and this feature being discovered in the Great Pyramid forces everyone to reexamine their theories. Theorists are once again challenged by new evidence and forced to accommodate the evidence or abandon their theories and conclusions regarding the construction of the Great Pyramid and its intended function. Of course, I was asked by several people what I thought about this find. I must admit I was a little taken aback by it. At the time it was announced, there was a great deal of uncertainty about the true shape of the void. The resolution provided by the muon scans did not reveal a well-defined shape with dimensionally precise walls, ceiling, and floor. What was revealed was a scattered point cloud that appeared random, and even the ScanPyramids team admitted there was not enough information to tell if the void was on the same angle as the Grand Gallery, above which it was located, or if it was horizontal. It was clear, however, that the void was very close to the North Shaft that runs from the King's Chamber to the outside of the pyramid. My initial reaction was that more information was needed before deciding whether it fit within the context of the power plant theory or not.

That information came sooner than I expected when I received a phone call from aerospace engineer Eric Wilson.

Eric is a graduate of Purdue University in Lafayette, Indiana, where he earned a master's degree in technology, and an undergraduate of Southern Illinois University, where he combined electrical and mechanical engineering in an advanced technical studies degree. I consider him to be a savant, in that his mind explores nature in ways that would make Tesla proud. In 2017, Eric contacted me after reading *The Giza Power Plant*, and as he lived only 80 miles away from me, we had lunch a few times and had long conversations about ancient technology and the pyramids. Eric had held some reservations about the maser theory I had proposed in the book, and now he called me while I was driving with Robert Vawter, who was traveling the country visiting friends and associates with his wife, Helen. (Unlike the output of a laser, a maser emits coherent microwaves, rather than coherent photons. When first invented, the laser was known as an optical maser.)

Eric had proposed that the schematic, which was based on previously discovered features inside the Great Pyramid and used when forming my analysis, was missing a preamplifier. Shortly thereafter, he learned about the Big Void in the Great Pyramid, its location near the North Shaft of the King's Chamber above the Grand Gallery, and that it appears to intersect or surround the shaft. Though more studies are needed to confirm whether the Big Void is in some way connected to the shaft, Eric explained that if it was what he thought it might be, it would resolve his previous reservations. What he had considered lacking was a preamplifier to amplify a hydrogen microwave signal before it entered the King's Chamber.

Discussing this further, we talked about what the void would need to have inside it to provide a boost to the signal. Eric said silicon quartz crystal and an electrical connection with the shaft would do the trick. He also pointed out some anomalies in the blocks near the location of the void that Gantenbrink had revealed in his drawing of the North Shaft. Further explorations in that area are necessary before drawing any conclusions about this introduction of new evidence, though it does stimulate optimism that the future will bring more enlightenment

ANATOMY OF AN ELECTRON HARVESTER

A   Pulse Generation Chamber

B   Hydrogen Generation Chamber

C   Acoustic Amplifier

D   Resonant Cavity Power Center

E   Chemical Feed Shafts

F   Input Signal Waveguide

G   Microwave Pre-Amplifier

H   Maser Power Output Waveguide

I   Anode

J   Feedback Control

K   Drain

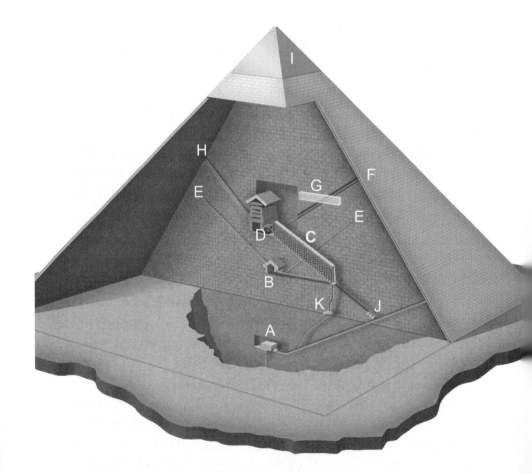

Figure 2.3. Anatomy of an electron harvester:
The Big Void "preamp."

about the Great Pyramid, especially now that some Egyptians are considering the possibility of it being a power plant.

As an aside, it is fascinating to me that cavities in the Great Pyramid that contain sand seem to crop up periodically. When I first traveled to Egypt in 1986, admittance to the Queen's Chamber was barred because a French team was boring holes in the Horizontal Passage. When the drill bit broke through into a cavity, sand poured out.

An Egyptologist who called in to the *Coast to Coast AM* radio show during my interview with George Noory on September 15, 2002, predicted that a 30-foot space filled with sacred sand would be discovered behind Gantenbrink's door at the upper end of the Queen's Chamber South Shaft. That is not what was discovered there, but what prompted her to make such a prediction? There may still be many mysteries and unseen features that could enhance our understanding of the technical function of this electron harvester, but when a large, previously unknown void in a strategic location can be seen as an integral part of the machine, how many other unknown features exist?

On September 28, 2021, I had a short meeting with Hany Helal and Hamada Anwar in 6th of October City near Giza. Both esteemed gentlemen are involved with the ScanPyramids mission. Helal is the former minister of higher education and scientific research in Egypt, and Anwar is an international judge. Helal was the mission's supervisor, and Anwar was Egypt's legal representative.

During the meeting I briefly described the contents of *The Giza Power Plan*t and gave them copies. I sensed a different reception than at the Cairo Museum in 1986 and felt encouraged to ask about the status of the ScanPyramids mission and what was planned for the future. They confirmed that further explorations were being planned but did not give details. In the meeting, I learned that Helal was an engineer specializing in rocks and mining. What better person to lead an exploration and study of the Great Pyramid? I began to recognize that a huge shift in attitudes had taken place in Egypt over the years.

Before we parted, Helal asked me a substantial question. I had previously stated that sometimes you can learn more from the questions asked than the answers given. In this case, what you learn can grow

Figure 2.4. The author (center) discussing the ScanPyramids mission with Hany Helal (left) and Hamada Anwar (right).

exponentially, depending on who is asking the question. The question was short, but the answer, or variety of answers, is long, and both form the basis for the final chapter in the book.

ANOTHER SCANNING PROJECT

True to the words of Helal and Anwar in September 2021, an article appeared on Phys.org on March 1, 2022, describing a new scanning project that was being planned by a different team to verify previous findings but also reveal with greater clarity and resolution other features that the ScanPyramids mission might have missed. The article reads in part:

> The Explore the Great Pyramid (EGP) mission uses muon tomography to take the next step in imaging the Great Pyramid. Like ScanPyramids before them, EGP will use muon tomography to image the structure's interior. But EGP says that their muon telescope system will be 100 times more powerful than previous muon

imaging. *"We plan to field a telescope system that has upwards of 100 times the sensitivity of the equipment that has recently been used at the Great Pyramid, will image muons from nearly all angles and will, for the first time, produce a true tomographic image of such a large structure,"* they write in the paper explaining the mission.

EGP will use very large telescope sensors moved around to different positions outside the Great Pyramid. The detectors will be assembled in temperature-controlled shipping containers for ease of transportation. Each unit will be 40 feet long, 8 feet wide, and 9.5 feet tall (12 m long, 2.4 m wide, and 2.9 m tall). Their simulations used two muon telescopes, and each telescope consists of four containers.

There are five critical points in the EGP mission:

- Produce a detailed analysis of the entire internal structure, which does not just differentiate between stone and air but can measure variations in density.
- Answer questions regarding construction techniques by being able to see relatively small structural discontinuities.
- The large size of the telescope system yields not only increased resolution, but enables fast collection of the data, which minimizes the required viewing time at the site. The EGP team anticipates a two-year viewing time.
- The telescope is very modular in nature. This makes it very easy to reconfigure and deploy at another site for future studies.
- From a technical perspective, the system being proposed uses technology that has been largely engineered and tested and presents a low-risk approach (Gough 2022).

The EGP team also promises to be able to gather data to create an accurate, tomographic, three-dimensional image of the pyramid by scanning it in slices, similar to how an MRI or CAT scan constructs an image of the human body. What this means, also, is that variations in material density will be defined, which is very exciting news and may put to rest the whole discussion about sand-filled cavities.

Don't expect results quickly, though. As noted, it is expected that the viewing time will be two years. However, as far as pyramid research projects have gone in the past, that is nothing to complain about, and in the meantime, the Great Pyramid will sit there patiently, as it has done for millennia, quietly humming its own unique melody.

THREE
PYRAMID PULSE

Everything in life is vibration.
ATTRIBUTED TO ALBERT EINSTEIN*

SINGING TO THE EARTH

For ancient Egypt's electron harvesters to work, vibration was essential. There is harmony in nature. From the cyclical frequencies of planetary systems to the frequencies of human DNA, vibration is omnipresent and influential in our lives. If it stops, we stop. That's the end. The symphony of humanity ceases to exist. Sound and music are at the core of our lives. Vibration moves us to action, startles us, soothes us, and provokes movement, whether the peaceful flow of meditation or the wild abandon of dance. When we are gathered in groups, sound entrains us with common purpose, such as with orchestral, choral, or dance performances—or even violent mob action. A single clap may stimulate the rest to join in applause.

The relationship between the dimensions of the Great Pyramid and those of the Earth is well documented and given extensive treatment in Chapter Eight of *The Giza Power Plant*, which relies on multiple

---

*The actual source of this quote is unknown. It fits many of Einstein's ideas but is not found verbatim in any of his works, despite frequent attribution.

sources to show that the Great Pyramid, dimensionally, is an integer of the Earth.

The granite complex inside the Great Pyramid, therefore, is poised ready to convert vibrations from the Earth into electricity. What is lacking is a sufficient amount of energy to drive the beams and activate the piezoelectric properties within them. (Now we can add the "Freund Effect.") The ancients, though, had anticipated the need for more energy than what would be collected only within the King's Chamber. They had determined that they needed to tap into the vibrations of the Earth over a larger area inside the pyramid and deliver that energy to the power center—the King's Chamber thereby substantially increasing the amplitude of the oscillations of the granite.

Modern concert halls are designed and built to interact with the instruments performing within. They are huge musical instruments in themselves. The Great Pyramid can be seen as a huge musical instrument with each architectural element designed to enhance the performance of the other.

While modern research into architectural acoustics might focus predominantly upon minimizing the reverberation effects of sound in enclosed spaces, there is reason to believe that the ancient pyramid builders were attempting to achieve the opposite. The Grand Gallery, which is considered to be an architectural masterpiece, is an enclosed space in which resonators were installed in the slots along the ledges that run the length of the gallery. As the Earth's vibration flowed through the Great Pyramid, the resonators converted the vibrational energy to airborne sound. By design, the angles and surfaces of the Grand Gallery walls and ceiling caused reflection of the sound and its focus into the King's Chamber. Although the King's Chamber also was responding to the energy flowing through the pyramid, much of the energy would flow past it. The specific design and utility of the Grand Gallery was to transfer the energy flowing through a larger area of the pyramid into the resonant King's Chamber. This sound was then focused into the granite resonating cavity at sufficient amplitude to drive the granite ceiling beams to oscillation. These beams, in turn, compelled the beams above them to resonate in harmonic sympathy. Thus, with the input of sound and the maximization of resonance, the entire granite complex, in effect, became a vibrating mass of energy (Dunn 1998, 160).

Understanding now that electrons in igneous rock within the Earth's mantle can be shaken loose and travel at great speed to the nearest highest point on the Earth's surface adds much more to the theory that the Great Pyramid was a power plant. It was not just a power plant relying on seismic mechanical energy to shake a granite chamber deep in the heart of a pyramid but a massive device that was designed and built with the accuracy of a machine. Yet it also functions like a musical instrument that not only utilizes vibration and the flow of sound within its passages and chambers but also controls the flow of electrons from within the Earth. From every square meter of stone covering 13 acres of limestone bedrock on a prominent plateau 10 miles west of Cairo, a vast amount of energy from the Earth was harvested to provide electricity to the civilization that designed and constructed the pyramids. See color plates 7 and 8.

But, you may ask, if we are relying on seismic energy to put pressure on igneous rock deep in the Earth to power these electron harvesters, won't we be waiting a long time? The answer is, we don't wait. We build into the electron harvesters the means to create their own mini-earthquakes. Geologist David Bressan writes for *Forbes:*

Tesla imagined using the waves generated by his invention for peaceful applications. One device would transform electricity into vibrations. Tesla then would use the rocks in the underground to send the vibrations to a second device. This receiving device would pick up the vibrations and transform the oscillations into electricity, to be used locally. In fact, so Tesla, [sic] the device, consisting of a piston vibrating in a cylinder, was already powerful enough to vibrate an entire building. Just one precaution was necessary. If powerful enough Tesla's machine could match Earth's frequency, causing even earthquakes. Still, in the 1930s, Tesla imagined using smaller devices to relieve energy from Earth, in this case, to *prevent* earthquakes. However, the "telegeodynamics" system by Tesla never managed to get beyond the prototype. The device was in reality not powerful enough to send energy far enough. Dampening of the oscillations by structures and the underground was far too strong. Another idea of Tesla was more successful. He imagined using the oscillations generated by his device to prospect the underground. Waves sent into the underground would be reflected by obstacles or different rocks. Observing the returning waves, a geologist may be able to see the underground. It's just this basic idea that modern seismologists use. Pulses of energy, generated by electromagnetic devices, controlled explosions or mechanical pistons, sent deep into the underground are reflected or deflected by geological structures. The reflected signals can then be used to reconstruct a model of the underground (Bressan 2020). [Italics added for emphasis.]

Becoming aware of Tesla's comments on earthquake prevention only recently, I did not credit him when I came to the same conclusion while writing *The Giza Power Plant:*

The technology that was used inside the Great Pyramid may be quite simple to understand but might be difficult to execute, even for our technologically "advanced" civilization. However, if anyone is inspired to pursue the theory presented in this book, their vision may be enhanced by the knowledge that re-creating this power source would be ecologically pleasing to those who have a concern about the environmental welfare and the future of the human race. Blending science and music, the ancient Egyptians had tuned their power plant to a natural harmonic of the Earth's vibration (predominantly a function of the tidal energy induced by the gravitational effect that the moon has on the Earth). Resonating to the life force of Mother Earth, the Great Pyramid of Giza quickened and focused her pulse, and transduced it into clean, plentiful energy.

Besides obvious benefits from such a power source, we also should consider the benefits that could be gained by utilizing such a machine in geologically unstable areas of the planet. As we discussed earlier, over time there is an enormous amount of this energy built up in the Earth. Eventually the weak spots in the mantle can give way to these stresses, releasing tremendously destructive forces. If we could build a device to draw mechanical energy from seismically active regions of the planet in a controlled fashion—instead of allowing it to accumulate to the destructive level of earthquakes— we might be able to save thousands of lives and billions of dollars. We would have a device that would help stabilize the planet; so rather than being periodically shaky real estate, California might eventually become the energy mecca of the United States, with a Great Pyramid drawing off the energy that is building up within the San Andreas Fault. A fanciful idea? Perhaps not. (Dunn 1998, 224)*

The concept of creating mini-earthquakes to draw seismic energy from the Earth in a controlled way to mitigate the occurrence of larger, more destructive earthquakes is not new, therefore. After mentions by Tesla in 1935 and myself in 1998, a different perspective was introduced in a paper published in 2014:

Laboratory experiments performed by different teams of the Institute of Physics of the Earth, Joint Institute for High Temperatures, and Research Station of Russian Academy of Sciences on observation of acoustic emission behavior of stressed rock samples during their processing by electric pulses demonstrated similar patterns—a burst of acoustic emission (formation of cracks) after application of current pulse to the sample. Based on the field and laboratory studies it was supposed that a new kind of earthquake triggering—electromagnetic initiation of weak seismic events has been observed, which may be used for the *man-made electromagnetic safe release of accumulated tectonic stresses* and, consequently, for earthquake hazard mitigation (Zeigarnik and Novikov 2014). [Italics added for emphasis.]

---

*With its prolific terrestrial light shows, Marfa, Texas might also yield a bumper crop of electrons.

Rather than using electromagnetic triggering to create these mini-earthquakes, in *The Giza Power Plant* I proposed that a mechanical device similar to Nikola Tesla's earthquake machine could be installed in the Subterranean Chamber that would periodically strike the ceiling, causing vibrations to flow through the pyramid.

How do we cause a mass of stone that weighs 5,273,834 tons to oscillate? It would seem an impossible task. Yet there was a man in recent history who claimed he could do just that! Nikola Tesla, a physicist and inventor with more than six hundred patents to his credit, one of them being the AC generator, created a device he called an earthquake machine. By applying vibration at the resonant frequency of a building, he claimed he could shake the building apart. In fact, it is reported that he had to turn his machine off before the building he was testing it in came down around him. *The New York World-Telegram* reported Tesla's comments from a news briefing at the hotel New Yorker on July 11, 1935:

"I was experimenting with vibrations. I had one of my machines going and I wanted to see if I could get it in tune with the vibration of the building. I put it up notch after notch. There was a peculiar cracking sound.

"I asked my assistants where did the sound come from. They did not know. I put the machine up a few more notches. There was a louder cracking sound. I knew I was approaching the vibration of the steel building. I pushed the machine a little higher.

Suddenly, all the heavy machinery in the place was flying around. I grabbed a hammer and broke the machine. The building would have been about our ears in another few minutes. Outside in the street there was pandemonium. The police and ambulances arrived. I told my assistants to say nothing. We told the police it must have been an earthquake. That's all they ever knew about it."

A reporter at that point asked Tesla what he would need to destroy the Empire State Building. Tesla replied:

"Five pounds of air pressure. If I attached the proper oscillating machine on a girder that is all the force I would need, five pounds.

Vibration will do anything. It would only be necessary to step up the vibration of the machine to fit the natural vibration of the building and the building would come crashing down. That's why soldiers break step crossing a bridge" (Pond and Baumgartner 1995, 5–6; Dunn 1998, 147–149).

CADMAN'S PULSE GENERATOR

While I didn't analyze the Subterranean Chamber in detail to see what kind of machinery could be installed and function in that manner, more details regarding the Subterranean Chamber became available during my tour in 2018, when two guests, geologist Adrian Lungan and engineer Andrew Leskowitz, explored the chamber during our two-hour private visit to the Great Pyramid. Before that, I had become very impressed with the work of marine engineer John Cadman of Bellingham, Washington.

Cadman had been emailing me information regarding his research into the Great Pyramid. He had studied and was inspired by the work of hydraulic engineer Edward J. Kunkel from Ohio, who had published his research on the idea that the Great Pyramid was originally built to serve as a ram pump that functioned through the force created by water descending through a long pipe. Kunkel's book, *The Pharaoh's Pump*, was first published in 1962 and captured the imaginations of numerous engineers and writers, including friends of mine of both stripes (Kunkel 1962).

Cadman built a model to Kunkel's specifications, but it did not function as he had described it would. Nonetheless, Cadman had the idea of scaling down his model and limiting it to the Descending Passage and the Subterranean Chamber.

He created a catchment basin on his hillside property near Bellingham, Washington, and, filling it with water, had a pipe feed water from the basin to a chamber he had made that was modeled after the Subterranean Chamber in the Great Pyramid. He added a waste gate that was, by scale, the same distance away from his model as the Great Pyramid is from the Sphinx enclosure, in the vicinity of which curious T-shaped openings are cut in the face of the rock that suggest they were connected with the pyramids on the plateau and were a part of the overall engineering scheme.

Having recently communicated with Cadman to discuss my reference to his work in this book, he shared with me by email some additional information: "The wastegate line should exit east of the Sphinx temple's midpoint–approximately 100' east and 30' below the surface. This is also the location of buried rose quartz granite that was discovered in 1980 by the Egyptian water department. This granite is not local to this area but came from 500 miles to the south (Cayce, Schwartzer, and Richards 1998, 145–146)" (John Cadman, email to author, May 30, 2022).

It has been reported by several researchers that the Giza Plateau is riddled with underground tunnels and deep shafts connecting them. Indeed, while walking around the pyramids on the Giza Plateau, you see several deep shafts covered with iron grates. Some shafts appear to be regularly shaped, with surfaces one might say appear as they did when they were first cut. Others appear to have suffered significant erosion, as though water had passed through them. The so-called Osiris Shaft under the causeway leading from the Sphinx enclosure wall to

*Figure 3.1. Shafts between the smaller Queen's Pyramids
to the east of the Great Pyramid.*

Khafre's pyramid is eroded in such a way. Gaining a complete under-standing of how these shafts and tunnels contributed to the overall operation of the pyramid complex is beyond the scope of this book and will have to rely on future studies by qualified scientists and engineers. Nonetheless, with evidence of shafts and tunnels in and around the pyramids, a connection with the openings that are evident in the cliff face near the Sphinx seems plausible, and it is not out of place to speculate that hydraulics, in some capacity, may have played a significant part in the function of the pyramids.

Cadman fitted the end of the pipe attached to his model of the Subterranean Chamber with a spring-loaded valve and submerged the assembly in a smaller catchment basin located downstream from the scale-modeled subterranean chamber. On opening the valve on the feed pipe and a riser pipe, water flowed into the chamber and a percentage of it was directed up the riser pipe that was attached to the top of the chamber. This served to push out any air trapped in the chamber. When water started gushing out of the top of the riser pipe, its valve was then shut, thereby directing all the water to the waste gate.

I was intrigued by what Cadman had accomplished, although it was a departure from my thinking on the subject, and I decided to accept an invitation to come to his property and witness the operation. When I thought about the thousands of hours independent researchers have spent over the years trying to figure out the secrets of the Great Pyramid, I recognized that John Cadman had invested not just time but money and sweat devoting his life to this research. In other words, he wasn't just an armchair theorist who believed that the Great Pyramid was not a tomb but had created something that actually worked! How could I ignore such devotion and untold hours of work? In November 2005, I booked a flight to Seattle and made my way by car to Bellingham to witness Cadman's pulse generator for myself.

John was an excellent and charming host, but I wasn't his only guest that day. Jack Kolle, PhD, a hydraulics engineer specializing in hydraulic pulse generators for the oil industry, had also been invited to witness Cadman's pulse generator in operation. After breakfast, served by John, we headed to his installation, armed with umbrellas and cameras.

*Figure 3.2. Left to right: John Cadman, Jack Kolle,
and Chris Dunn at Cadman's pump.
Image credit: John Cadman.*

I was very surprised when Cadman started up his pulse generator. Something I wasn't quite expecting happened. The pipe leading to the waste gate was buried under the path, and as I was walking along the path, a slow steady beat shook the ground. However, it wasn't an ordinary beat. Its timing was 60 beats per minute, but more significantly, it beat like a heart. Lub dub . . . lub dub . . . lub dub. I was happy I had taken the time to experience this, as its effect was more profound than any distant analysis could produce.

Kolle seemed impressed. I asked him what he thought was creating the heartbeat effect. He claimed that he had no idea, but hydraulics could sometimes be considered to be a "black art," where results are achieved through trial and error, and that no computer modeling software was capable of predicting such outcomes. I then asked him if he thought what we were witnessing was the design intention of the builders of the Great Pyramid. Cautiously, he said no.

Cadman has continued to refine and build his pump and has selected another site on his property to locate the chamber inside a large gap in a giant boulder. With his brother Mike, another excellent host who provided me with lodging for two nights when I returned in 2016, they have improved the site with a visitors center and even restrooms. In Bellingham, the Cadman passion still runs strong. Besides the Great Pyramid, John Cadman's other passion is breeding Black Russian Terriers. (His website for Sentinel Kennels provides information on his terriers, while Research Article 41 on his website provides information on his pyramid pulse generator research.)

With respect to the availability of evidence within the Great Pyramid that supports Cadman's theory, there are features of his model that have still not been discovered. A connection to the waste gate from the Subterranean Chamber has not been confirmed. When I first witnessed Cadman's model in operation, I thought that there could be a continuation of the horizontal channel that is cut in the south wall of the chamber. However, two guests who joined my Lost Technology tour in 2018, Andy Leskowitz and Eric Wilson, separately crawled down this channel and reported it was a dead end with no sign of another channel going anywhere. Moreover, both said that there didn't appear to be any signs of an opening being blocked off.

The only other place where a connection might exist is the deep well shaft in the chamber floor. Richard William Howard-Vyse was a Lieutenant-Colonel when first visiting Egypt in 1835 and was promoted to Colonel in 1837, when he joined engineer John Shae Perring in exploring the pyramids. Howard-Vyse's campaign in the 1830s revealed many previously unknown features about the Great Pyramid. While a great deal of attention has been given to his discovery of the so-called Relieving Chambers above the King's Chamber, Peter Tompkins provides a lengthy though significant caption underneath a drawing of the Subterranean Chamber that reports on Howard-Vyse's activities there:

> The subterranean pit cut deep into the bedrock is almost 600 feet directly below the apex of the Pyramid. It is 31 feet (9.45 m) in the east-west direction, but only 27 feet (8.23 m) north-south.

Though its ceiling is relatively smooth, its floor is cut in several rough levels, the lowest being 11 feet 6 inches (3.50 m) from the ceiling.

In the south wall, opposite the entrance, is a low passage that runs another 53 feet (16.15 m) southward before coming to a blind end.

In the center of the floor is a square hole, which was 12 feet (3.66 m) deep in 1838, but was dug deeper by the English explorer Howard-Vyse in the vain hope of finding an outlet for a further hidden chamber. (Tompkins 1971, 9)

Leskowitz and Wilson confirmed Howard-Vyse's account of the horizontal shaft being a dead end, and Howard-Vyse appears to have already revealed that the pit cut into the floor has no connection to any shaft that would serve as a conduit to a waste gate outside the pyramid—despite not being satisfied with reaching its termination point and proceeding to dig further down into the bedrock. Another blow to the hydraulic pulse generator theory came from geologist Adrian Lungan, who reported that he didn't see any signs in the Subterranean

Figure 3.3. Ceiling of the Subterranean Chamber.
Image credit: Andrew Leskowitz.

Figure 3.4. Subterranean Chamber pit.
Image credit: Andrew Leskowitz.

Chamber of the continuous movement of water. Leskowitz confirmed his observations.

I'm disappointed that Cadman's model has not held up under the scrutiny of existing evidence. I wanted him to be correct, but just as I have been forced to accept and accommodate the mistakes I have made in my analyses, hopefully Cadman can revise his theory and produce another model that works without the proposed waste gate's function. In a recent email to me, he said that this is what his plans are: "Also, the 'dead end shaft' will be built to be a dead end of correct length. (I'll cap the [dead-end] shaft of the ink jet model. It didn't need tuning.) This was discovered after doing sound analysis last 2 years. My focus is the vortex of the sub chamber. My inspiration is Theodore Fujita's work with his tornado simulator at Chicago University. The future pulsing ink jet model is going to be as effective as Theodore Fujita's model. They designed perfect water saturated vortex for a reason" (John Cadman email received on May 30, 2022).

Obviously, Cadman's proposal that a hydraulic pulse generator was a part of an electron harvester energy system is not off the table—quite the contrary. The question is, though, was it used in the Great Pyramid? If not, what are we left with—a mechanical device that delivers timed hammer blows to the ceiling of the Chamber? Tesla's earthquake machine?

In considering that idea, could the square depression that exists in the ceiling of the chamber be the result of constant pounding like that of a manufacturing press? Or, if Eric Wilson's large-scale acoustic engine, described in Appendix C below, which only needs a single, large mechanical blow to cause it to run continuously until interrupted is a possibility—and he has assured me that a proprietary experimental machine has been demonstrated to function this way—then the question becomes, "Am I overthinking this particular critical part of the electron harvester?" The answer is, I don't know. Possibly, but I have to continue on the path with Tesla.

Figure 3.5. Nikola Tesla (1856–1943) at age 40.

When viewed with consideration of a Tesla-like oscillator, could the features found in the Subterranean Chamber be considered as the ground preparation for housing such a machine? With that in mind, was it a mechanical device that was installed in the pit and activated by a hydraulic, pneumatic, electromagnetic, or steam-driven piston housed in a cylinder installed in the 53-foot-long tunnel located in the south wall? There has to be some logical reason for boring out a long tunnel to nowhere. Perhaps the machinery was delivering alternating hammer blows to the ceiling, to vibrate the pyramid, and the bottom of the pit, to couple the pyramid with the Earth. Then again, perhaps it provides, as Wilson describes his machine, that necessary single blow to initiate action until interrupted and called upon again?

LIVING IN HARMONY
WITH YOUR NEIGHBORS

Another consideration is that the demand for electricity may have dictated the frequency and force of the action in the Subterranean Chamber, thereby increasing or lowering the pyramid's power output. Also, we cannot ignore the Great Pyramid's lovely neighbors!

Today, large power plants provide a base load to the power grid. When more electricity is called for, smaller, gas-fired peaking power plants spring into operation to satisfy that demand. With respect to the smaller pyramids on the Giza Plateau, I wrote: "The three smaller pyramids on the east side of the Great Pyramid may have been used to assist the Great Pyramid in achieving resonance, much like today we use smaller gasoline engines to start large diesel engines" (Dunn 1998, 219).

Today I believe that the reverse may be closer to the truth. That is, the Great Pyramid stimulated output from the smaller pyramids. The electrons rushing to the surface of the Earth from deep in its mantle would not all miraculously seek out just the Great Pyramid before entering the atmosphere. Before deciding where to locate the pyramids, a study may have been performed to discover where the highest con-centration of electrons naturally escapes the Earth when stimulated to move. This would be analogous to prospecting for gold, though the yield is more valuable and inexhaustible.

A complete survey of the many deep shafts and tunnels on the Giza Plateau may provide answers to the questions these speculations create, and we may learn that the progenitor of John Cadman's hydraulic pulse generator system that caused the ground to shake lies not within the Great Pyramid, or any other pyramid for that matter, but elsewhere, and deeper underground. Created before the pyramids were built, its purpose may have been to serve all the pyramids on the plateau by shaking electrons loose from igneous rock.

It could be that the pulses generated in the Great Pyramid's Subterranean Chamber are specific to just the Great Pyramid and its performance as a maser. The big daddy of pulse generators that causes the entire plateau to resonate may not be in any of the pyramids but deep under the bedrock of the Giza Plateau, functioning as Cadman's research and model have demonstrated.

THE HUMAN ELEMENT

Knowing that all life is vibration and that vibration we know as sound affects humans, I was intrigued to receive an email from musician and composer Susan Alexjander. In September 2000, we exchanged several emails, and through those, and information on her website, Our Sound Universe, I learned about her collaboration with Professor David Deamer of the University of California, who had mapped the frequencies of human DNA. Alexjander had stepped down the DNA frequencies to human hearing range and used the result to compose music. As I had referred to Tom Danley's work in 1996–1997 when he conducted sound tests inside the Great Pyramid, Alexjander made the connection between Deamer's results and Danley's, which is that F# is a dominant frequency in both human DNA and the King's Chamber in the Great Pyramid. Susan was kind enough to send me *Sequencia*, a most enjoyable CD of her compositions based on Deamer's human DNA frequency results.

Danley's results received wide recognition when, in 1997, one of the producers of a documentary being made during his work in the Great Pyramid appeared on Art Bell's *Coast to Coast AM* radio show. Producer Boris Said reported Danley had discovered that the King's Chamber had a natural resonance approximating an F# chord between 2 and 20 hertz in the infrasonic range. He also reported that the King's Chamber floor was sitting on nodes of an eggcrate-like structure—suggesting that the ancient architects were intentionally trying to achieve an acoustical device where damping of any vibration in the floor is minimized. This is also supported by the fact that the granite chamber is not connected, or tied in, to the limestone-core masonry, having a space surrounding the entire assembly, and the walls are not pressing down on the floor but resting on limestone-core masonry five inches (12.7 cm) below the floor level.

Danley's long-awaited article about his personal experience during the expedition appeared in the July/August 2000 edition of Live Sound International.

Asking the question "How Loud Does It Get?" Danley reported:

While doing an FFT on the between-sweep time or quiet parts of the recording I found some very LF sound—resonances which start at a few Hz and go upward to 15–20 Hz or so. At least some of these were the same LF resonances I excited with my sweep, but not all of them. This sound was present even if everyone is silent.

I crunched the results of the measurements, and they were sent on to a musicologist that was part of the staff. As mentioned, he identified that there was a pattern of frequencies, which roughly form an F-sharp chord.

Not all the resonances fell in the right place but many did and some repeated the pattern for many octaves. In other words, it was roughly tuned to F-sharp over many octaves.

It has been suggested (by others) that the Great Pyramid is NOT a tomb at all but actually a temple of sorts and that these resonant frequencies were "designed into" the structure. While many exotic and often far-fetched properties have been ascribed to "the power of the pyramid," I see a possible argument that some of the phenomena people experience in it may be caused by the acoustical properties that were measured.

The effects of LF sound were extensively studied by various government agencies to determine the effects on humans, partly for the space program. One of the things that was discovered is that infrasound (very LF) can effect [sic] ones brain wave activity (Alpha rhythms, etc.) and other biological functions.

If, as some suggest, these pyramids were constructed as a "temple" or for an initiation ritual rather than a tomb, then the LF sounds may be deliberate and have served a sacred purpose—with the sound triggering and even forcing changes in brain wave state (i.e., one's level of consciousness) (Danley 2000).

BACK TO THE FREUND EFFECT

The human connection to vibration and how we are affected by certain frequencies became a subject of interest for Friedemann Freund and his research into preearthquake processes. It is well-known that animals exhibit abnormal behavior prior to major earthquakes. In their article titled "Unusual Animal Behavior Prior to Earthquakes: A Survey in North-West California," David Jay Brown and Rupert Sheldrake write:

> Observations of unusual animal behavior prior to earthquakes have been reported around the world since the beginning of recorded history (Tributsch 1982). In particular, the Chinese and Japanese have recorded these observations for many hundreds of years (Lee, Ando, and Kautz 1976) and have made attempts to incorporate these reports into an earthquake warning system with some success (Allen 1976). For example, on February 4, 1975, the Chinese evacuated the city of Haicheng several hours before a 7.3 magnitude earthquake largely on the basis of unusual animal behavior observations (Allen 1976).
>
> The anomalous behaviors most frequently reported include restlessness or excitability, a heightened sensitivity to mild stimulation,

vocal responses, a tendency for burrowing, premature termination of hibernation, and leaving their normal habitats. The precursory lead times vary from just a few seconds to more than several months (Lee, Ando, and Kautz 1976). These unusual behaviors have been reported in a wide diversity of animal species, including many varieties of mammals, birds, reptiles, fish, and insects (Tributsch 1982). (Brown and Sheldrake n.d.)

Freund's interest in this subject was not limited to how animals were affected by impending earthquakes but also included whether evidence exists to suggest that humans also became affected. In a 2014 lecture to the Society for Scientific Exploration, Freund presented some startling information:

Here is something that just blew me away when I first read it. I was asked to review a paper that was first published in a book by the Russian author Alexander Shitov. In 2003 there was a large earthquake, magnitude 7.6 in a sparsely populated part of southern Siberia at the boundary between Russia and Mongolia, an area with a population of 250,000 people. After the earthquake, Shitov collected every piece of information he could find to identify what had happened before this major event.

After receiving the medical records for the whole province, he found that the number of people who visited the emergency rooms had tripled starting about 10 days before the earthquake. Obviously, at that time, nobody knew the earthquake was coming, but people began to flock to the emergency rooms, complaining of various forms of discomfort and actual illnesses. Some had pneumonia, others exhibited phenomena linked to the central nervous system such as epileptic attacks, hypertension and other phenomena that clearly pointed to the functioning of the central nervous system. And so, we are starting to understand what is happening between ultra-low frequency electromagnetic waves and electron transfer processes in the single cell. (Freund 2014) Note: Transcript edited by Friedemann Freund for clarity.

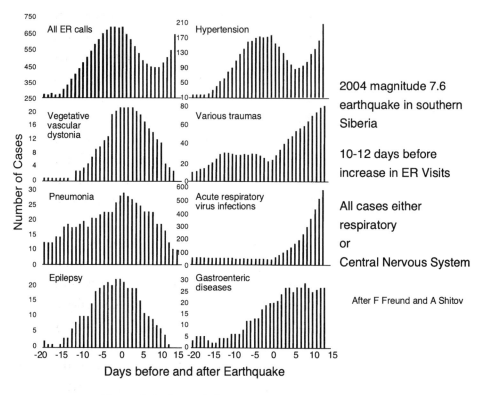

Figure 3.6. Recorded emergency room visits prior,
during, and after an earthquake.

ANSWERS TO A CENTURIES-OLD MYSTERY

In *The Giza Power Plant*, I questioned the forces at play behind how I was physically affected while visiting the Great Pyramid and spending time in the King's Chamber. At the time I wrote it, I thought that skeptics would attack me for it, and I was right. However, I felt it was significant enough to include. I had no idea then that 25 years later I would be armed with answers from a surprising variety of sources and would address it again in a new book that expands on what I wrote in 1998, not just on this subject but in every aspect possible:

Covering a large land area, the Great Pyramid is, in fact, in harmonic resonance with the vibration of the Earth—a structure that could act as an acoustical horn for collecting, channeling, and/or focusing terrestrial vibration. We are led to consider, therefore, that energy associated with the pyramid shape is not drawn from the air or magically generated simply by the geometric form of a pyramid, but that the pyramid acts as a receiver of energy from within the Earth itself. It could be, also, that these infrasonic sound waves provide an explanation for the physical phenomena some people have felt when entering the Great Pyramid. The "pyramid energy" that has inspired countless numbers of people since the time of Napoleon may be the effects of infrasonic sound on the brain, which is said to resonate at around 6 hertz.

I experienced the phenomenon myself while in Egypt in 1986. After being inside the Great Pyramid for about an hour, I found myself in a rather uncomfortable situation. Sick and in immediate need of a bathroom, I really did not know if I was going to make it; but I rushed out of the King's Chamber, down the Grand Gallery and Ascending Passage, squeezing past tourists. Once outside I ran down the hill to the Mena House and headed straight for the bathroom. I made it just in time. The bathroom walls in the Mena House were constructed of Aswan granite. As I relaxed and closed my eyes, and without any external influence, the resonance of the King's Chamber filled my head. At the same time a pyramid shape began to glow in the center of my forehead. It was only after leaving the Mena House that this sensation faded. It could be that this phenomenon is only felt at certain times, according to the seismic activities within the Earth. I have not experienced it since then, although I have been inside the Great Pyramid several times since (Dunn 1998, 144).

I would add that infrasonic sound affects not just the brain but the entire human organism, which reacts to seismically induced energies, or the natural rhythms of the Earth, passing through the pyramid. These include infrasound as well as the Freund Effect, with the King's Chamber becoming the focal point of these energies. The combination of effects described by Freund and Danley and the Russian research team led me to suggest that the Great Pyramid was a receiver of these energies and still is, albeit not at the same level as when it was functioning as an electron harvester. I don't know how much clearer the evidence can be to scientifically explain what has previously been interpreted to be a "mystery."

Unrelated to the Egyptian pyramids, this phenomenon has not gone unnoticed by other cultures over the years and has even influenced the selection of locations for holy, or sacred, places of worship and meditation. In "Earthquake Lights in Legends Of The Greek Orthodoxy," Igor V. Florinsky wrote:

> Local legends may contain information about real geological events of the past. Earthquake lights (EQL) can occur in the atmosphere over earthquake epicenter areas and adjacent faults before and during quakes. They may look like diffuse airglow, flashes, fiery pillars, and luminous balls. EQL may cause a mystical experience probably due to the influence of their electromagnetic fields on the brain. Subjective perception and interpretation of EQL depend on religious and cultural traditions. We study a stereotype of EQL interpretation in the legends of the Greek Orthodoxy exemplified by the legends about the foundation of two shrines: St. George Monastery near the Cape Fiolent, Crimea and the Panagia Tripiti Church in Aigion, Peloponnese. It is argued that the similar interpretation of EQL observation in the Crimean and Peloponnesian legends were caused by similar natural and economic living conditions of the Greek population in the both regions in the Middle Ages. We also consider some examples of EQL observation took place in other regions and their interpretation in other denominations. Differences and relations between EQL of mechanoelectrical and degassing origin are discussed. Finally, we consider the role of active faults in the production of a mystical experience and sacralization of an affected landscape (Florinsky 2016, 159).

Mystical experience is a sort of altered state of consciousness characterized by visual and auditory hallucinations, including apparent contacts with divine or supernatural creatures. These contacts are often accompanied by extreme emotions, such as delight, euphoria, horror, panic, etc. (Florinsky 2016, 160).

While the mystery of the pyramids has attracted millions of visitors over the years and is promoted not only by New Age groups but also by leading minds in Egyptology, removing the mystery and providing a solid foundation of understanding based on physics and science can only attract more. The energies of the Earth will continue to be amplified in the Great Pyramid regardless of what we may believe about their nature. Tour operators will continue to promote their tours to draw visitors who are attracted to esoteric, occult, and seemingly magical experiences. Some do, or will, believe that the purpose for building the pyramids was to provide those who enter them a place where a deeper connection to the consciousness of the universe may be felt and may refer to them as initiation chambers.* But those who have not visited the pyramids and have not enjoyed that experience, when understanding that there is an explanation based on physics and science, will be able to understand that it's not all due to an overactive imagination or "in the head" of the human being affected but that the source for this stimulation is in the Earth and in the pyramid itself.

Am I goring a sacred cow? Some may argue that I am and choose to ignore what decades of dedicated research by qualified physicists like Friedemann Freund have produced. Regardless of what I think and write here, nobody can predict how they will be affected while being inside the Great Pyramid or whether a companion who joins them in the King's Chamber will experience the same effect. Maybe they will not be affected at all. After visiting Egypt 14 times, I have learned that being moved by an unusual experience inside the Great Pyramid during one visit does not mean that the next visit will give me the same experience. This, I believe, is because of fluctuation in the Earth's activities causing a higher level of seismic activity on one visit rather than another. I also wonder whether the time of the year that you visit affects these fluctuations.

---

*These effects are discussed in *Awakening the Planetary Mind* by Barbara Hand Clow. (Clow 2011, 83–88)

Just the concept of an energy source that is environmentally friendly and is in harmony with nature is inspiring in and of itself without needing to enter the Great Pyramid. Becoming inspired and elated is probably not the experience we would anticipate having when exploring the energy center of a defunct, abandoned, and decaying modern power plant.

Wanting to know how my Facebook friends who have visited the Great Pyramid were affected within its confines, I asked for feedback. Some preferred to share privately, while others publicly gave detailed accounts of their experience. A few of their responses were:

- Feeling as though they were inside a powerful machine.
- Started to get light-headed, then while sitting felt like there was a ball of energy above me.
- Had a miraculous healing experience in the Subterranean Chamber.
- Had a miraculous healing experience in the King's Chamber.
- In the coffer and intoning I felt as though my voice was powering a huge subwoofer and felt like I was lifting out of my body.
- I was there alone and felt a strong presence.

One of the responders to my question answered privately because they did not want to publicly share. I understand that because, historically, many have had what they thought were supernatural experiences but don't talk about them for fear of being ridiculed. One of my friends shared the following in a private message:

Just wanted to tell you about my experience in the King's Chamber. I was in it when we started doing the chants . . . the resonance was shaking me and was actually frightening. If I tried to join in, the resonance was so loud I couldn't continue.

I tried to relax and catch my breath, and suddenly the lights all went out. The blackness was something I'd never experienced before, it was quiet and calm and I was finally feeling calm and I seemed to see lights before my eyes. They were still there whether my eyes were open or not. It was very serene once I relaxed but there was definitely some energy entering my body.

I'm not a real spiritual person but it was quite profound and will stay with me forever. I felt like I was connecting to something powerful.

I spent about 15 minutes in "the box"* quite an experience.

Napoleon Bonaparte is probably the most famous person in history who is reported to have exited the pyramid a bit shaken up but close-mouthed about what he experienced inside. However, this legend might be just a myth with no accurate records to support it. In "Napoleon at the Pyramids: Myth Versus Fact," historical fiction writer Shannon Selin provides reason to question this hoary legend. She writes:

> The third myth appears to be of more recent vintage. Napoleon is said to have spent some time alone in the Great Pyramid, from which he emerged visibly shaken by some experience that he refused to divulge to anyone else.

Napoleon Stayed Outside

The problem with these legends is that no one who was with Napoleon in Egypt reports that he ever entered a pyramid. According to Napoleon's private secretary, Bourrienne:

"On the 14th of July Bonaparte left Cairo for the pyramids. He intended spending three or four days in examining the ruins of the ancient Necropolis of Memphis; but he was suddenly obliged to alter his plan. This journey to the pyramids, occasioned by the course of war, has given an opportunity for the invention of a little piece of romance. Some ingenious people have related that Bonaparte gave audiences to the mufti and ulemas, and that, on entering one of the great pyramids he cried out, 'Glory to Allah! God only is God, and Mahomet is his prophet!' Now, the fact is, that Bonaparte never even entered the great pyramid. He never had any thought of entering it. I certainly should have accompanied him had he done so, for I never quitted his side a single moment in the desert. He caused some persons to enter into one of the great pyramids while he remained

---

*King's Chamber granite sarcophagus.

outside, and received from them, on their return, an account of what they had seen. In other words, they informed him there was nothing to be seen" (de Bourrienne 18381, 208; Selin 2017).

I'm not sure how the Napoleon rumor began, when it began, or who started it, but it has been reported quite convincingly by many, including Tompkins, who wrote:

> Meanwhile Napoleon, whose logistical mind enabled him to figure that the Great Pyramid and its Giza neighbors contained enough stone to build a wall 3 meters high and one meter thick all around France, had become attracted by the arcane qualities of the King's Chamber.
>
> On the twenty-fifth of Thermidor (the Revolutionaries' August 12, 1799) the General-in-Chief visited the Pyramid with the Imam Muhammed as his guide; at a certain point Bonaparte asked to be left alone in the King's Chamber, as Alexander the Great was reported to have done before him. Coming out, the general is said to have been very pale and impressed. When an aide asked him in a jocular tone if he had witnessed anything mysterious, Bonaparte replied abruptly that he had no comment, adding in a gentler voice that he never wanted the incident mentioned again.
>
> Many years later, when he was emperor, Napoleon continued to refuse to speak of this strange occurrence in presage of his destiny. At St. Helena, just before the end, he seems to have been on the point of confiding to Las Cases, but instead shook his head saying, "No. What's the use. You'd never believe me" (Tompkins 1971, 49–50). (Note: Tompkins provides no in-text footnotes or citations.)

It appears that, according to Tompkins, Napoleon was just as concerned about the reaction to his sharing what he had experienced inside the Great Pyramid as my Facebook friend. But Napoleon's, my friend's, and my own shared accounts of the unusual effects that are experienced by so many people who enter this energy machine do not count as evidence to support the arguments put forth in this book. These anecdotes, while interesting, are separate from the scientific and experimental reports that have been performed.

Comparing the cultural differences in society today with those of 100 years ago, there is no doubt that human interactions with technological advances have significantly affected how we live, work, and play. Those born today grow up surrounded with symbols of progress, from the architecture that fills major cities to the houses they live in and the transportation they use to get from one place to another. With these influences around them, they connect with a utility that allows them, in Star Trek fashion, to reach across the world and talk by videophone to people they know personally and those they have just met virtually. In writing this book, I have benefitted from the ability to ask a question of a friend in Egypt and receive a response in less than an hour. If I suddenly lost that ability, I would have to adjust, but I would probably adjust more peacefully than someone who has never experienced the world before these technologies became available.

> *I don't think there is such a thing as a real prophet. You can never predict the future. We know why now, of course; chaos theory, which I got very interested in, shows you can never predict the future.*
> ARTHUR C. CLARKE (CLARKE 1968)

Because we were not born into an environment in which the pyramids functioned as electron harvesters with associated technologies, engendering a culture that was influenced by and reacted to these technologies, we can only ask: What did that civilization look like? How did the people live? What abilities did they possess—their own natural abilities and those connected with and facilitated by the technologies in their possession? We can imagine, but in truth, our imaginations will most likely fail us.

A case in point is the introduction of robots in the manufacturing workplace. In the 1970s, computers were becoming more powerful, and manufacturing machines were modified to operate automatically. One article I read in a trade magazine in the '70s contained a prediction that robots would eventually replace highly skilled machinists and toolmakers, which, along with other reasons, resulted in young people being guided into different careers because of a perceived redundancy in the future. The article discussing these developments included a graphic that showed a human-like robot operating a manual lathe. In much the same way as a human does, the robot's hands were gripping and turning the wheels that caused movement of the lathe's axes. The concept the writer had of how robots would replace machinists in the future, it turns out, could not be further from the truth. The technology built into the robot that allowed it to stand on its robot legs, pressing buttons, flipping handles, and turning wheels with its robot hands, was installed in the lathe itself. Such are the limits we possess in accurately predicting how new technologies will change our environment at work or at home.

The great classical master composers, I believe, were accessing information in the universe that is profoundly unique and beyond the reach of normal people. Perhaps they were drawing from the same source described by Tesla as "the core." I ask the question: "What was going on in their brains when they were creating such masterpieces?" What was their dream state like? It is as though a section of the universe opened up and spilled into their consciousness the beauty of heaven in the form of an organized, disciplined order of musical notes. Did they visualize this order as it played in their heads prior to being set down on paper? Did they hear it? How can you describe something you have had no previous experience with? We get to experience and enjoy the results of a process that remains a closed book.

With the testimony of many visitors to the Giza Plateau who speak of what has been described as a supernatural experience within and around the Great Pyramid, a reasonable framework based on physics provides us with an understanding that natural forces, rather than supernatural magical sources, are at work there, which again brings us back to the famous quote by Arthur C. Clarke: *"Any sufficiently advanced technology is indistinguishable from magic."*

In the case of the Great Pyramid Electron Harvester, those powerful pulses that were crunching igneous rock and providing unimaginable benefits to the society that built it are now pulsing quietly and causing *hum*ans to hum.

FOUR

THE KING'S ORCHESTRA

THE MUSIC OF THE PYRAMIDS

Deep inside the beating heart of a mountain of stone is a finely tuned musical instrument that was designed to draw into its space acoustic energy from other finely tuned instruments that were balanced and positioned precisely in a row, ready to convert mechanical vibrations into airborne sound. Manufactured and assembled with machine-like precision, the King's Chamber and Grand Gallery orchestra worked in unison and harmony, and while most of the public's attention for the past 20-plus years has been focused on the Queen's Chamber and what lies at the end of an 8 inch (20.32 cm) square shaft, the musical ensemble above it is infinitely more magnificent, more complex in design and construction, and, from a purely scientific and engineering perspective, sufficiently advanced technologically that many who have experienced being within its space have left with the feeling that they were in the presence of something supernatural. As if they were surrounded by magic.

Although they perform different functions as parts of this electron harvester, the King's Chamber and Grand Gallery cannot be addressed separately without one recognizing the other. Little happens in the King's Chamber without the Grand Gallery functioning as it was designed, and without the King's Chamber, the Grand Gallery is unnecessary.

If one considers the Great Pyramid to be a tomb, it would be reasonable to ask why this supposed resting place for a king would have most

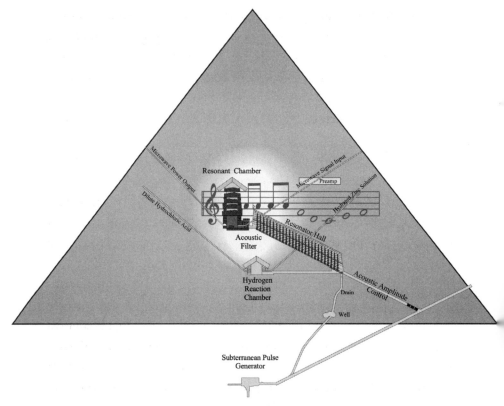

Figure 4.1. The King's Chamber and orchestra. (See also color plate 9.)

of its parts hidden from view? Didn't the king deserve to appreciate the work that had been done on his behalf? The first person to penetrate the barriers that prevented all those who had entered the pyramid before him from beholding the enormous labor and fortune was the British Colonel Richard William Howard-Vyse. In 1836, thousands of years following this extraordinary expenditure of time and labor, Howard-Vyse, while conducting explorations in the Great Pyramid, was in a crouched space above the King's Chamber, examining a mysterious layer of granite beams above him that were similar to the granite beams that formed the ceiling of the King's Chamber beneath him. The space is named "Davison's Chamber" after Nathaniel Davison, who had discovered it in 1765.

Howard-Vyse, who had received £10,000 from his family to conduct this exploration and, reportedly, relieve themselves from his presence, was intent on making a significant discovery and, so far, was not having any luck. The granite layer above his head posed a tantalizing clue that something might lie beyond. Noticing a crack between the beams, Howard-Vyse considered the possibility of yet another chamber above. Being able to push a three-foot-long reed into the crack without obstruction seemed a clear indication that there must be another space beyond.

Abandoning the slow and tedious results that hammers and chisels produced, and armed with a supply of gunpowder and hashish, Howard-Vyse recruited one of his workers to risk life and limb by setting charges and scurrying off while limestone and granite were blasted away and showered down the vertical shaft that was slowly inching its way up above the King's Chamber until another chamber was revealed.

Similar to Davison's Chamber, a ceiling of monolithic granite beams spanned the newly discovered chamber, indicating to Howard-Vyse the possible existence of yet another chamber above. After blasting upward for three-and-a-half months and to a height of 40 feet (12.19 m), they discovered three more chambers, making a total of five. To construct these five chambers, the ancient Egyptians found it necessary to use 43 pieces of granite weighing up to 70 tons each. The red granite beams are cut square and parallel on three sides but were left seemingly untouched on the top surface, which was rough and uneven. Some of them even had holes gouged into their top sides.

KHUFU'S CARTOUCHE

When Howard-Vyse's team blasted through to reveal the uppermost chamber, they discovered that it was fitted with massive gabled limestone blocks, one of which was adorned with a cartouche that contained within it the name of Khufu. There has been quite a bit of controversy surrounding the validity of these markings, with some conspiracy theorists claiming that Howard-Vyse himself had painted them so he could claim a major discovery. Are they original to the time the pyramid was built, or were they added later? I have been asked my opinion of them on several occasions over the years, and while I do not take sides on who is correct in the debate about the origins of the markings, my opinion has always been that they are graffiti, whether they were painted when the pyramid was built, or later, when Howard-Vyse discovered them. From an engineering perspective, graffiti is irrelevant when trying to understand the design and function of an engineered product. An example would be the graffiti painted on the side of a railroad car has nothing to do with the purpose for which the railroad car was built. Having said that, I recognize that to historians these marks may have value and may stimulate a much wider and deeper investigation regarding their origins that doesn't involve engineering.

Along that same line of thinking, young people today may recognize the name Tesla as being a manufacturer of electric automobiles. They may learn, after being told this, that Tesla was a genius who died 80 years ago after ushering in the electrical age by inventing alternating current (AC) transmission. There is no question that the car maker and the man have no connection, other than an honorary recognition of the man's pioneering efforts by the modern electric car manufacturer. In our small window of time in the life of the planet, there is no mistaking the application of the name Tesla on products with the actual human. Can we say the same for humans in the future? Living as we are within this time frame, this span of time is significant. When considering prehistory and the age and purpose of what was created thousands of years ago, a hundred years is a mere sliver on the time chart.

Considering the enormous effort that must have gone into delivering to the Giza Plateau the granite monoliths above which these inscribed gabled blocks are situated, it's natural to wonder why all this work was necessary? By today's standards, the task of quarrying just one of the granite beams at the Aswan quarries and transporting it 500 miles down the Nile to the Giza Plateau before raising it approximately 200 feet (60.96 m) in the center of the pyramid would not be a simple or inexpensive task. That this step was repeated 43 times surely gives us a hint that these monoliths are not just construction materials, even though, to the ancient Egyptians, these monoliths are a fraction of the size of other monoliths that had been and were in the process of being extracted from the quarry. The ancient Egyptians didn't seem to be daunted by what was necessary to create obelisks and statues that weighed in excess of 1,000 tons out of hard, igneous rock.

If Howard-Vyse had contained his enthusiasm within the limits of modern archaeological methods and standards, chances are we would not know that these enigmatic chambers exist. But we do, and we have him to thank for it. However, they need to be explained in a reasonable and scientific way. Howard-Vyse surmised that the reason for the five superimposed chambers was to relieve the flat ceiling of the King's Chamber from the weight of thousands of tons of masonry above and allow it to have a flat ceiling, rather than a gabled ceiling similar to the ceiling in the Queen's Chamber. This explanation seemed to have stuck and is repeated in mainstream literature. However, as I argued in *The Giza Power Plant*, the 43 giant granite beams that spanned 27 feet (8.23 m) in length and weighed up to 70 tons are holding up no more than their own weight, and the structures responsible for holding up the blocks of the pyramid above are the limestone gable blocks in the uppermost chamber. I argued that the Queen's Chamber could have been fitted with a flat ceiling without the multiple layers of beams above. I thought my reasoning was logical and didn't expect any opposing argument.

I was wrong. The internet and social media can always produce an opposing argument for anything in the world. As you can imagine, arguments against the Giza power plant theory have been prolific and creative. One such argument against my assertion that the five layers of granite beams above the King's Chamber were redundant came to my inbox by way of South African Egyptology student Mikey Brass, who alerted me to an article that had appeared in a German magazine the year following the publication of my book, where the question was discussed with researcher Frank Dörnenburg. Dörnenburg writes:

I have been debating elsewhere, the King's Chamber and the question of why five "relieving" Chambers were needed to be used to spread the massive weight above the King's Chamber. My answer to this was I simply did not know.

A good answer to this question can be found in Göttinger

Miszellen 173/1999, p 139 ff: Ludwig, Daniela; Offene Fragastellungen in Zusammenhang mit der Cheopspyramide in Giza aus baukonstruktiver Sicht.

"The old method of corbeling channeled the weight force directly to the walls of the chamber. The new, and here for the first time used, gable-roof redirects the force down AND sideways. If the Egyptians had put the gable roof in the King's Chamber directly on the ceiling, like in the Queen's Chamber, the sideways force would have damaged the Great Gallery. So they had to put the gable above the upper layer of the Galleries construction. The easiest way to do this is to stack small chambers. And if you look at the cross section you will see, that now the sideways force of the roof goes well over the roof of the gallery."

Superficially, what is proposed here may seem plausible. It is, however, a construct founded on flawed assumptions and an incomplete analysis of the entire King's Chamber complex. Consider the following.

The hypothesis assumes that dynamic lateral forces would follow the direction of the angled blocks and that these lateral forces would accumulate as more stone was piled on top of the gabled bocks. According to the hypothesis, the consequence of each block added above the King's Chamber causes additional lateral thrust to push against the southern end of the Grand Gallery.

The drawing in Figure 4.2 represents a mechanical setup with which many manufacturing technologists would be familiar. It is a steel plate resting in a V-block. If we allow that the above hypothesis is correct, the plate will push on surface A, causing lateral movement.

Such is not the case. At rest, the plate will put more pressure on the opposite surface B, due to the center of gravity of the piece. There are no dynamic forces at work. There is only dead weight, which is distributed according to each member's center of gravity. When an object is placed on an inclined plane, it has the potential to move down that plane by gravitational forces acting upon it. This movement continues until an obstruction is encountered, at which time the lateral motion ceases.

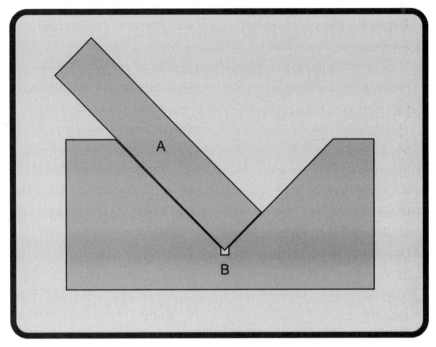

Figure 4.2. V-block holding rectangular plate.

The gabled ceiling blocks above the King's Chamber are situated on an inclined plane cut into the core blocks. Assuming that, like the Queen's Chamber, the center of gravity of these blocks lies outside the chamber walls, the blocks may be described as cantilevered, whereas there is no arch thrust at the apex where two opposing blocks meet. The entire weight of the block is borne by the core masonry blocks that have an inclined plane upon which the gable block sits, with some weight being carried by the block that captures the lower end.

With the Grand Gallery's southern wall blocks butted against its east and west wall blocks, any lateral forces from the King's Chamber's gabled ceiling blocks give less cause for concern than, say, the forces acting on the roof of the Horizontal Passage from the pressure of the Queen's Chamber's gabled ceiling blocks or the pressure of the blocks bearing down on the roof of the Grand Gallery. Moreover, building on top of gabled ceiling blocks does not necessarily mean they must bear a tremendous accumulation of weight. As described in Figure 4.3, the distribution of weight does not necessarily have to bear down on the gable. A core masonry block can be sized and placed so that the gable blocks have little to no weight on them at all.

Without knowing for sure what design features were employed, I can envision a sound design that would not damage the Grand Gallery.

The argument that the builders were directing the lateral forces of the gable blocks away from the Grand Gallery to protect it from movement is based on a drawing of the King's Chamber and Grand Gallery looking from east to west. But when the King's Chamber and Grand Gallery is viewed from north to south, the argument dramatically falls apart. A precise arrangement of large multiton blocks can ensure that their weight will be directed vertically and will not impact the Grand Gallery or place undue pressure on the King's Chamber's gabled blocks. Figure 4.3 illustrates the Grand Gallery's indifference to the King's Chamber's gabled blocks, should they be used to cover just the ceiling of the King's Chamber.

When seen from north to south, it becomes clear that structurally, the assembly of monolithic granite beams above the King's Chamber,

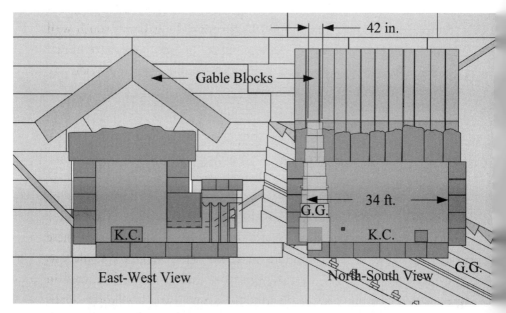

Figure 4.3. East-west and north-south view of King's Chamber.

with the protective gabled limestone above, has no influence on the Grand Gallery. The top of the Grand Gallery is only 42 inches (1.07 m) wide, while the widest point at the bottom is 82 inches (2.08 m). The King's Chamber is just over 34 feet (10.36 m) long—over nine times the width of the top of the Grand Gallery next to the ceiling. Even if we accept that there would be undue pressure on the south wall of the Grand Gallery, which I don't, it would not necessitate five layers being built across the entire length of the King's Chamber. Also, why five layers of beams? Why not a large open space with the gabled ceiling above?

In cutting these giant monoliths, the builder evidently found it necessary to craft the beams destined for the uppermost chamber with the same precision and finish as those directly covering the King's Chamber. Each beam was cut flat and square on three sides, with the top side seemingly untouched. This is significant, considering that the beams directly above the King's Chamber would be the only ones visible to those entering the pyramid.

The rough measurement between the ends of the gabled blocks and the Grand Gallery south wall is about 9 feet (2.74 m). Considering the width of the Gallery (between 42 and 82 inches [1.07–2.08 m]), it is reasonable to assume that the blocks forming the gallery's south wall extend to the east and west outside the inner surface. The conventional argument for the so-called Relieving Chambers does not provide any relief when searching for the truth.

KING'S AND QUEEN'S CHAMBERS NORTH SHAFT TRICKERY

More evidence inside the Great Pyramid that escaped my attention when I wrote *The Giza Power Plant* was the location and design of the King's and Queen's Chamber North Shafts. It is generally accepted that these shafts change direction to avoid interference with the Grand Gallery as they get closer to it. As recorded by earlier explorers and illustrated in great detail by Rudolf Gantenbrink, the King's Chamber North Shaft, like the Queen's Chamber North Shaft, contains several bends, which

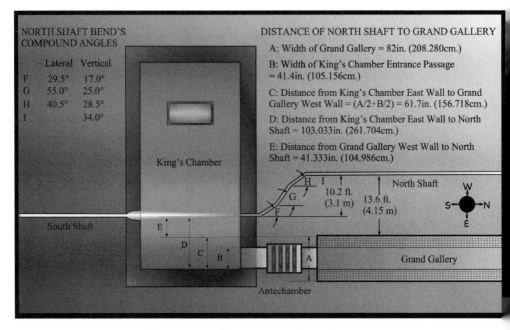

Figure 4.4. Plan view of King's Chamber, Grand Gallery,
and North Shaft. (See also color plate 10.)

allows a greater distance between the shaft and the Grand Gallery. That
is a reasonable assumption and one I tend to agree with. However, the
question is, by how much does the shaft need to stay clear of the gallery?

What is really mind-blowing and worth remembering is that both
the King's and Queen's Chamber North Shafts *could have been* constructed, like their South Shafts, without bends, because they are located
3.4 feet (1.04 m) from the Grand Gallery's west wall. I overlooked that
detail when I wrote *The Giza Power Plant*, but the extra difficulty the
builders undertook to guide the shaft further than 3.4 feet away from
the Grand Gallery's inside wall is a significant detail that cannot be
ignored, and it contributes to the overall picture of the extraordinary
effort expended by the builders and how they did not do anything by
chance or on a whim. See Figure 4.4 and color plate 10 for the measurements and angles of the shafts.

Considering that the King's Chamber North Shaft bends around the
Grand Gallery strongly suggests the blocks that form the gallery walls

had to be thicker than 3.4 feet (1.04 m). Why do the Grand Gallery wall blocks need to be thicker than 3.4 feet (1.04 m)? In the context of the electron harvester theory, there is a logical answer to this question. Like the King's Chamber, the Grand Gallery does not rely on the surrounding core masonry for its structural integrity, with the exception of the masonry underneath that is holding it up. This is because its function relies on greater stability and precision than the core masonry that surrounds it. It is built to withstand continuous vibration generated in the Subterranean Chamber. However, unlike Tesla's apartment building that was threatening to shake apart from the harmonic forces imparted by his earthquake machine, the Grand Gallery did what it was designed to do: pass the vibrations through to resonators suspended at intervals along its length and convert them to airborne sound. The movement of that sound relies on fixed and precise dimensions.

NOT TRICKERY—PHYSICS AND ENGINEERING

While the question of why the North Shafts need to be a substantial distance away from the Grand Gallery is being considered, a more important and fundamental question has been overlooked—and was missed by me when I wrote *The Giza Power Plant*. The question is, why do the shafts need to be at this exact location? If they were air shafts, or shafts through which the King's soul would pass, surely the openings could be anywhere in the chamber—further to the west, east, closer to the ceiling or the floor? But they are not, and considering the extra work involved in placing them in that specific location, there has to be a good reason for it.

The answer to this question is critically related to the function of the pyramid machine. As Dustin Carr, PhD, first pointed out to me, and Robert Vawter and Eric Wilson later confirmed, these shafts appear one quarter of the distance along the length of the chamber.

This is significant, for that location is where the highest amplitude (energy) exists in a standing wave in a resonant cavity. Moreover, it is where the input of energy will have the most effect.

Figure 4.5 is an isometric view of the King's Chamber North Shaft and its relationship with the King's Chamber and Grand Gallery. The

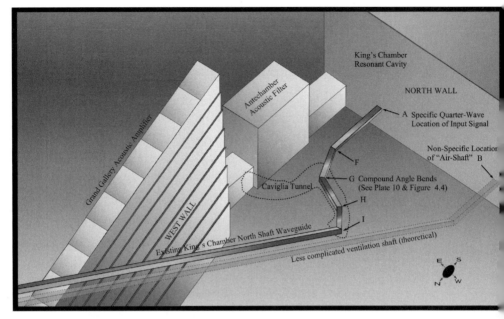

Figure 4.5. Isometric view of King's Chamber, Grand Gallery, and North Shaft. (See also color plate 11.)

three-dimensional geometry was collected by Rudolf Gantenbrink and was once available for viewing through a CAD viewer on his website, Cheops. The website is now archived and available at the UPUAUT Project page on the ISIDA Project's website. Gantenbrink's excellent work allows us to visualize these shafts more clearly and punctuate the points that follow:

- Shaft A is the existing shaft, which has multiple lateral bends that redirect it to/from the quarter-wave location in the King's Chamber. These redirections occur three-dimensionally and the vertical angle decreases with each turn (see Figure 4.4).
- Shaft A could redirect with just two bends. But it has four. Why?
- Shaft B is a theoretical air shaft that only has one bend.

If the purpose of these shafts was to ventilate the chambers, a simple, straight, horizontal shaft would suffice.

CAVIGLIA'S TUNNEL

Caviglia's tunnel, as illustrated in Figure 4.5, was excavated in the mid-1830s by Giovanni Battista Caviglia, who was hired by Colonel Richard William Howard-Vyse to join him in blasting through areas of the Great Pyramid in his hunt for treasure. The entrance, which is fitted with an iron gate and normally padlocked, is located in the short horizontal passage between the Grand Gallery and the Antechamber. During one of my many visits, I noticed the padlock had been removed, so I entered the tunnel and made my way to the North Shaft. It was just a brief visit while using a rather dim flashlight, but I was able to notice that the upper section of the shaft was left exposed up to where Gantenbrink had installed a ventilation fan.

WEIGHING THE EVIDENCE

There is only one published theory that provides answers to the questions raised by the existence of the King's Chamber North Shaft. This was published in *The Giza Power Plant:*

The North Shaft served as a waveguide through which the input microwave signal traveled. A typical waveguide is rectangular in shape, with its width being the wavelength of microwave energy and its height measuring approximately half its width. The North Shaft waveguide was constructed precisely to pass through the masonry from the north face of the pyramid and into the King's Chamber. That microwave signal could have been collected off the outer surface of the Great Pyramid and directed into the waveguide.

The (originally smooth) surfaces on the outside of the Great Pyramid are dish-shaped and may have been treated to serve as a collector of radio waves in the microwave region that are constantly bombarding the Earth from the universe. Amazingly, this waveguide leading to the chamber has dimensions that closely approximate the wavelength of microwave energy 1,420,405,751.786 hertz (cycles per second). This is the frequency of energy emitted by atomic hydrogen in the universe. (Dunn 1998, 185–186)

The wavelength of hydrogen is 8.309 inches. The width of the King's Chamber North Shaft is 8.4 inches.

Now, after examining the shaft's details more closely and considering the newly discovered Big Void as well as the advice of a brilliant physicist and aerospace engineer, we can add more evidence to the list that was published in *The Giza Power Plant*. See color plates 12 and 13.

- North Shaft dimensions approximate those necessary for the design of a waveguide.
- The South Shaft has a bulbous opening that is similar to a microwave horn antenna.
- Extra care and knowledge were necessary to include bends in the North Shaft so that it entered the chamber at the exact quarter-wave location.
- Instead of a two-bend transition to reach the quarter-wave point, the North Shaft has four. While the design of waveguides is a specialty, with craftsmen carefully tuning them using various means, Eric Wilson was not surprised to see four bends instead of two, for, as he explained to me, they would facilitate the direction and power of the beam.
- Wilson, who had studied Gantenbrink's CAD drawings very closely, also noted that transition points were included along the upper length of the shaft. A channel block further up the shaft, which is longer than the others, is seen to have a radial protrusion in the ceiling, while another area close to the bottom transitions from a larger height to a smaller one. These features, he said, are not uncommon when designing and building an extremely long waveguide, which can be something of a "black art" that requires unique skills and knowledge.
- We also cannot forget the potential relationship of the North Shaft with the recently discovered Big Void, which, like the North Shaft, appears to be located to the west of the Grand Gallery

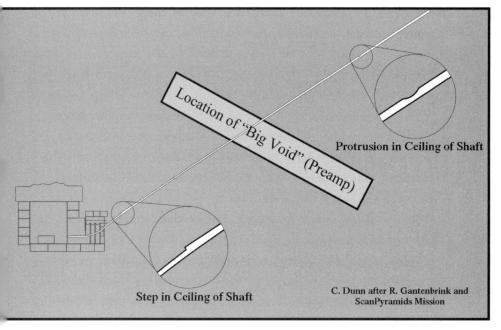

Figure 4.6. Anomalous feature in the North Shaft.

halfway between two anomalies that are noted above and also shown in Figure 4.6. Whether this observation is significant is yet to be determined. Further research promises to be more accurate in its results.

There doesn't appear to be any reason other than a waveguide for the King's Chamber North Shaft to exist. It's a highly technical and complicated inclusion in the pyramid that required skill and precision to assemble. It cannot be dismissed lightly or assigned a mundane non-technical function.

WHEN THE EARTH SHOOK

When the Grand Gallery was built and its resonators assembled inside, it functioned as designed without relying on frequent maintenance or tuning.

Other features and what can be described as mysteries exist within the King's Chamber and Grand Gallery. Many of them found potential explanations in *The Giza Power Plant*. One explanation related to Sir William Flinders Petrie's observations of the condition of the chamber itself.

> The King's Chamber was more completely measured than any other part of the pyramid; the distances of the walls apart, their verticality in each corner, the course heights, and the levels, were completely observed. On every side the joints of the stones have separated, and the whole chamber is shaken larger. By examining the joints all-round the second course, the sum of the estimated openings is 3 joints opened on N. side, total = .19 ; 1 joint on E. = .14 ; 5 joints opened on S. = .41 ; 2 joints on W. = .38. And these quantities must be deducted from the measure, in order to get the true original lengths of the chamber. I also observed, in measuring the top near the W., that the width from N. to S. is lengthened .3 by a crack at the S. side.
>
> These openings or cracks are but the milder signs of the great injury that the whole chamber had sustained, probably by an earthquake, when every roof beam was broken across near the south side; and since which the whole of the granite ceiling (weighing some 400 tons) is upheld solely by sticking and thrusting. Not only has this wreck overtaken the chamber itself, but in every one of the spaces above it are the massive roof beams either cracked across or torn out of the wall, more or less, at the south side; and the great eastern and western wall of limestone, between and independent of which, the whole of these construction chambers are built, have sunk bodily. (Petrie 1883, 27)

The King's Chamber/Grand Gallery assembly is a most magnificent piece of evidence that presents more awkward questions than the enigmatic King's and Queen's Chamber shafts. Within the King's Chamber sits a granite box, which has been referred to as a sarcophagus by Egyptologists. The dimensions of the box indicate it was installed in the chamber before the chamber had been completed, as its dimensions are larger than the low passageway that is the only access to the chamber.

The color of the box has been reported as being chocolate brown. It has also been severely damaged, with the southeast corner broken at the top. I had proposed in *The Giza Power Plant* that the displacement of the King's Chamber walls and the cracked beams above were not caused by an earthquake but by forces within the chamber itself, following the introduction of oxygen into a highly energized hydrogen atmosphere that resulted in an explosion. To support my contention, I referred to the lack of disturbance and precision of the lower passages and chambers as well as the base upon which sit the pyramid stones that covered 13 acres of limestone bedrock with no greater variation in level than ⅞ inch (2.22 cm).

My theory was related to several pieces of evidence. The first is the King's Chamber North Shaft, which had dimensions that would be suitable for use as a waveguide. Because the width of the shaft is 8.4 inches (21.34 cm) and the wavelength of hydrogen is 8.309 inches (21.104 cm), I proposed that hydrogen was created in the Queen's Chamber below and filled the open spaces of the pyramid. I will address, in the following chapters, more evidence revealed since that reinforces this theory. The second piece of evidence is the damage to the chamber and beams above. The third is the granite coffer's difference in color and its damaged corner.

In *The Giza Power Plant* I also posited the theory that the top surface of each granite beam above the King's Chamber, rather than receiving less attention than the flat, square sides and bottom surfaces, were crafted with more attention but from a different class of worker.

> Rather than suffering from a lack of attention, therefore, the rough top surfaces of those granite beams in the King's Chamber have been given more careful and deliberate attention and work than the beams' sides or bottoms. Before the ancient craftspeople placed them inside the Great Pyramid, each beam may have been "tested" or "tuned" by being suspended on each end in the same position that it would have once it was placed inside the pyramid (see Figure 4.7). The workers would then shape and gouge the topside of each beam in order to tune it before it was permanently positioned inside the

Figure 4.7. The tuning of a single granite beam can be accomplished
by suspending it between two blocks at the ends, striking it,
and selectively removing material from the top side until it "rings"
at the correct frequency.

pyramid. After cutting three sides square and true to each other, the remaining side could have been cut and shaped until it reached a specific resonating frequency. The removal of material on the upper side of the beam would take into consideration the elasticity of the beam, as a variation in elasticity might result in more material being removed at one point along the beam's length than at another. The fact that the beams above the King's Chamber are all shapes and sizes would support this speculation. In some of the granite beams, I would not be surprised if we found holes gouged out of the granite as the tuners worked on trouble spots. What we find in the King's Chamber, then, may be thousands of tons of granite that were precisely tuned to resonate in harmony with the fundamental frequency of the Earth and the pyramid! (Dunn 1998, 157)

A granite beam can be cut flat on the top to achieve a particular frequency when struck. It might be that its uneven top surface was cut not only to achieve a specific resonant frequency but also to equalize the release of positive hole electrons across the length of the beam, and to do that, they needed to avoid an uneven distribution of peroxy defects in the minerals. This may be considered merely speculation, nevertheless, a similar condition exists in neodymium-doped yttrium aluminum garnet (Nd.YAG) laser rods, which I became familiar with in the 1980s when processing aircraft engine parts.

While neodymium atoms in a laser rod, when excited by light, produce an infrared 1064 nm wavelength photon, peroxy defects in igneous rocks, when excited by movement, produce positive hole electron propagation. Where and in what amount of concentration they exist within their crystalline substrates is determined after the crystal is formed.

In discussing this with Robert Vawter, he introduced the idea that the movement of sound through the granite stack might be affected by the uneven distribution of materials in the granite, principally quartz, felspar, and mica, and that this could affect the coherency of sound as it exits a layer of granite beams, travels through open space, and impacts the beam above. All of this is highly speculative, of course. But then, until each hypothesis is proven or not proven through the application of scientific principles and physics, what isn't?

Smyth and Petrie unwittingly provided clues that this resonance theory may not only be plausible but indeed probable. Both sought an explanation for the holes gouged near the ends of these granite beams. Smyth said, *"These markings, moreover, have only been discovered in those dark holes or hollows, the so-called 'chambers,' but much rather 'hollows of construction,' broken into by Colonel Howard-Vyse above the 'King's Chamber' of the Great Pyramid. There, also, you see other traces of the steps of mere practical work, such as the 'bat-holes' in the stones, by which the heavy blocks were doubtless lifted to their places, and everything is left perfectly rough"* (Smyth 1880, 7). Rather than seeing them as holes used for lifting the blocks into place, Petrie speculated on an alternate reason for Smyth's so-called 'bat-holes': *"The flooring of the top chamber has large holes in it, evidently to hold the butt ends of beams which supported the sloping roof-blocks during the building."* (Petrie 1883, 31)

Neither Smyth's nor Petrie's explanations are particularly satisfactory. The most likely and logical reason for the holes gouged near the end of the beam may have been to strategically weaken the beam in order for it to respond more readily to sound input.

According to Boris Said (pronounced Sigh-eed), who was with engineer Tom Danley when he conducted his acoustical tests inside the King's Chamber, the King's Chamber's granite beams resonated at a fundamental frequency and the entire structure of the chamber reinforced this frequency by producing dominant frequencies that created an F-sharp chord. Not surprisingly, the F-sharp chord is believed to be in harmony with the Earth. While testing for frequency, Danley placed accelerometers in the spaces above the King's Chamber, but I do not know whether he went as far as checking the frequency of each beam. Boris Said shared something in his interview with Art Bell that may be some indication of where Danley was heading with his research. He said that the beams above the King's Chamber were "like baffles in a speaker." Further research would need to be conducted before any assertion could be made as to the relationship these holes may have with tuning these beams to a specific frequency. However, when we consider the characteristics of the entire granite complex, along with other features found in the Great Pyramid, it seems clear that the results of this research will be along the lines of what I am theorizing. (Dunn 1998, 157) (Note: Unfortunately, as of today, this extensive and highly specialized research has not taken place.)

When I visited Egypt in 1986, I had already become convinced that acoustics were at play in this gigantic instrument of musical ingenuity. In 1995, I was able to privately make a recording in the King's Chamber, and I presented my observations and results in *The Giza Power Plant*. With some corrections and modifications to shorten the text, here is the essence of what I wrote:

I recently had the opportunity to confirm the acoustical phenomena of the King's Chamber in a unique and rather fortuitous way, although without Danley's instrumentation or expertise. On February 24, 1995, I paid the inspector of the Giza Plateau $100 to leave me inside the Great Pyramid after all the tourists had left and it was officially closed . . . . My backpack was weighted with some instruments I had brought along specifically to take some acoustical and electromagnetic frequency measurements . . . . I had asked for the lights to be turned off because I did not want any background electrical noise to affect the digital frequency counter with which I was equipped. I had brought this along to measure radio frequencies that I believe can be generated by the resonant chamber inside the Great Pyramid. I also had brought along a tape recorder, which I turned on and placed upon a block of granite situated close to the granite coffer, or "box" in the King's Chamber. Using this block as a work stage, I positioned my flashlight, the digital frequency counter, and a monochromatic tuner, which measures sound frequency and is used to tune musical instruments. The last batch of tourists made their way down the Ascending Passage to the outside.

When the noise faded away, I began to test the frequency of the granite box. I had read in a booklet by flautist Paul Horn (that accompanied his album *Inside the Great Pyramid*) that the granite box resonated at a frequency of 438 cycles per second (Hz). Horn had used a Korg tuner, which is quite a bit more expensive than the Matrix tuner I was using, to perform his own acoustical tests inside the chamber. I thumped the side of the box with my fist. The tuner registered between 439 and 440 Hz. I loudly hummed that note to test the resonance of the King's Chamber. Sliding up the scale, I noted that the reverberation faded until I produced the same note one octave higher, and then the reverberation was even greater. It was then time to test for radio frequencies, and the lights were still not turned off. I was becoming concerned because I had to be out of there in 15 minutes, so I crouched back through the passageway, stood at the top of the Grand Gallery, and yelled to the guards to turn off the lights. While there, I intoned the note I had hummed earlier, then scurried back into the King's Chamber with the intent of being by my flashlight before I was pitched into darkness. At last, the hum of the lights ended abruptly and I was left with only the light from my flashlight. My heart was pounding in my chest. I was sweating profusely. I had reached the moment of truth.

Then I realized that the hum of the lights was masking the whir of fans installed in the alleged "air channels." The guards had turned the lights off, but, as I discovered later, they could not turn off the fans without a key to the electrical box that controlled them. Disappointed that I would not be able to use any data, I went ahead and took some readings anyway.

Back in my hotel room I played the tape back and discovered three very interesting acoustical phenomena. First, the tape had picked up overtones to the note I was humming in the King's Chamber. I had been unable to hear them while I was in the chamber because I was the source of the sound. This was a very exciting discovery, as it supported my theory on the King's Chamber. I tempered my enthusiasm, however, with the thought that it might be the equipment that was resonating at a higher frequency and generating this overtone. Mine was not a very expensive tape recorder, but still I believe the overtones were generated by the chamber complex itself (including the granite ceiling beams) because of the King's Chamber's true design and purpose. The second discovery revealed that when I was on the Great Step at the top of the Grand Gallery yelling to have the lights turned off and then humming the tone, the playback sounded as if I had never left the room. At that moment, except for a small passageway, there were thousands of tons of granite and limestone that separated me from the recorder, and my voice was being projected into a 28-foot-high and 157-foot-long expanse of space. The third revelation was that the footsteps and noises I made, while traversing the low passage to and from the Grand Gallery, reverberated in the King's Chamber and caused it to resonate at its natural frequency. This was noted on the Matrix monochromatic tuner that was turned on as I played back the tape. My footsteps and noises were registering at approximately 440 Hz. I particularly noted that the fluctuations of the tuner during this section of the playback were not as great as those I had observed inside the King's Chamber when I was humming the frequency (Dunn 1998, 140–143).

Robert Vawter

In February 1997, I learned from researcher Stephen Mehler about a sound engineer named Robert Vawter who had done some studio analysis of a tape recording that he (Mehler) had made inside the King's Chamber. As he related it to me, those results were the same as mine. I was given Vawter's phone number and promptly gave him a call. Vawter confirmed what Mehler told me. He said that he had digitally processed the tape provided by Mehler, and he was able to isolate harmonic overtones of the intoned frequency. Vawter claimed that the King's Chamber was designed specifically as a resonant chamber in which the sound of specific frequencies would resonate. He said that every dimensional feature of the chamber he had studied indicated the manifestation and form of harmonic resonance. (Dunn 1998, 143)

In recent conversations, Vawter told me that he was inspired to gather acoustic data from the pyramids by Abd'el Hakim Awyan, a well-known personality and guide who lived in the village of Nazlet El Samman, which is located next to the Great Sphinx. Of particular interest was their discussion about King Sneferu's two pyramids at Dahshur. Awyan's interpretation of Sneferu is that the name means Double Harmony. He also referred to the pyramids as Per Neters and interpreted that to mean Places of Nature. Vawter not only provided an analysis of Mehler's recording, but also collected acoustical data from Sneferu's Red Pyramid. Since his studies in the '90s, Vawter has expanded his research into pyramid acoustics, and in Appendix A he has kindly provided a deeper understanding of his thorough research, which encompasses not just pyramid acoustics but significant observations regarding ancient acoustics in general.

Dustin Carr

On June 12, 2003, I received an email from physicist Dustin Carr:

Dear Mr. Dunn,

I am impressed with your work on reverse engineering the great pyramid. I am in strong agreement with you that there was a purpose for the pyramids, and power generation is the only thing that would justify the size of the structure. I have a Ph.D. in physics from Cornell, 2000, and have worked as a lead researcher/manager at Bell Labs. If you search the web for "nanoguitar" and "lambda router" you will find some reference to my work.

I have some ideas that I think would go far in helping you refine the theory, but I need a little more data. There are some types of "engines" that were not considered in your book, but I need to know the precise dimensions of the various parts; the kings chamber, the lengths of the shafts, the sarcophagus. I am an expert in vibration physics, and have access to certain tools that could be of benefit in the acoustical analysis.

I hope to hear from you soon.

Dustin W. Carr, PhD

(Email received 6/12/2003 9:37:16 AM)

Naturally, I was extremely happy and privileged to receive Carr's offer of help. I communicated with him and his colleague, Stephen Turner, PhD, also of Cornell, off and on for several months. In this time, I learned that Carr's preliminary calculations using the geometry and dimensions of the King's Chamber and its shafts confirmed my original claim that the ancient Egyptians' design intent was acoustics. However, it was more complicated than I had imagined. The design intent was not the application of simple acoustics but reflected a sophisticated knowledge of nonlinear acoustics, of which at the time of our communication in 2003 we still had a lot to learn.

Carr reached a point in his analysis when an official research proposal was needed so that on-site measurements and acoustic samples could be taken. For many reasons, this failed to materialize and a tantalizing question mark exists in its place. Robert Vawter's analysis, as described in Appendix A, reinforces that question mark, and we both hope that the information in this book will provide impetus for qualified people with, as Professor Sherbini appropriately pointed out, organizational weight behind them to take up the challenge and provide solid answers to our questions.

I learned through recent communication with Carr that he was unable to be involved in this research at this time, though he told me I was welcome to share his previous studies and comments.

WHEN THE MUSIC STOPPED

Throughout the Earth's history, tranquility has frequently been violently interrupted. Earthquakes, volcanos, and associated devastation continue unabated to frequently remind us that the Earth is a dynamic planet and what might exist on its surface through the ingenuity and labors of humans can be wiped out in an instant. Less frequent than the general localized damage from the Earth's internal forces are the occasional interruptions visited by external forces that affect the entire planet. Asteroids, meteorites, and coronal mass ejections from the sun are known to have impacted the Earth in the past, causing apocalyptic global destruction.

While the body of knowledge that describe these events in detail is beyond the scope of this book, I accept that cataclysms have occurred in the past and that one particular cataclysm reduced a large section of the civilization existing at the time to ruin. While I am not able to specifically identify which one, I can confidently say that much of what we now recognize as advanced technology in the ancient world is the result of human activity that was violently interrupted by outside forces that brought everything to a halt. While Egyptologists have surmised that the Great Pyramid at Giza was built in the Fourth Dynasty, or about 4,500 years ago, strong arguments are made to consider that the pyramids were constructed thousands of years earlier, before a cataclysm brought them and their creators to ruin.

The damage to the pyramids and other magnificent examples of engineering genius crafted in igneous rock was so great that the survivors had no hope of restoring their world to the level of accomplishment they previously enjoyed. Such would be our own fate should a similar event happen in the future.

Taking all this into consideration, I argued that the most likely force that pushed out the walls of the King's Chamber and caused the beams above to crack would have been a hydrogen explosion inside the chamber that probably followed a cataclysmic event that overcame the Earth and destroyed most of the existing civilization. With the pyramids being finely tuned to resonate with the Earth's natural frequencies, the introduction of an outside force that impacted the Earth was too much for the pyramids to handle. They started to shake apart. Casing stones were shaken off. Some, like the pyramid at Meidum, have casing stones piled up around the base. Other structures, like the Sun Temple at Abu Ghorab, have large limestone blocks that appear to have been ejected from their original position, with at least one I have seen having cracks running through at awkward angles, leaving one to wonder how and why they are so situated. Some, such as the pyramid at Abu Rawash, were under construction and the work was not completed.

The Unfinished Obelisk at Aswan, I believe, was not abandoned because it had developed a crack. Considering the enormous amount of work that went into ploughing out a deep channel around it, a lot of useful granite had been exposed that could still have been used, even if not for an obelisk. Like every other work in progress at the time the cataclysm hit, everything came to a violent end.

EVIDENCE OF CONFLAGRATION
IN THE GREAT PYRAMID

In *The Giza Power Plant*, the evidence used to support the explosion hypothesis was the use of hydrogen suggested by the dimension of the North Shaft, the localized damage to the chamber, the coffer, which, as I surmised, was cooked, and the theory that these devastations were caused by a cataclysmic event that affected not just the Great Pyramid but all structures throughout Egypt. I was not aware when the book was published in 1998 that other evidence existed in the Grand Gallery to support this theory but was hidden beneath a layer of dirt and dust.

Recognizing that the Grand Gallery was not in pristine condition, as it would have been after it was first built, I addressed what possible effects a violent explosion in the King's Chamber may have had on this critical section of the machine.

Figure 4.8.
Resonator assembly.

The awesome force unleashed inside the King's Chamber—of such magnitude that it melted granite—also would have consumed other susceptible materials. If the resonators in the Grand Gallery were made of combustible material, such as wood, they most likely would have been destroyed at the same time. Evidence to support this speculation comes from reports that the limestone walls in the Grand Gallery were subjected to heat and, as a result, the limestone blocks calcinated or burned. The disaster that struck the King's Chamber, therefore, may have been responsible for destroying the resonators. (Dunn 1998, 212)

There is no information that informs us what shape and out of what material the proposed resonators were made. While I previously discussed the use of wood, this was based on its natural ability to resonate and contribute to the performance of a musical instrument. In considering this material for use so close to the energy center in this electron harvester, I was concerned about the level of heat the resonators would be exposed to under normal operating conditions. My current conclusion is that it is still an open question. However, whatever material the resonators were made of succumbed to the hydrogen explosion.

My illustration of the resonators has not changed much since the publication of *The Giza Power Plant.* That is because whatever I might change it to would be just as speculative as what I previously imagined, which was based on the dimensions and features of the gallery. This included the use of the 27 pairs of slots that are spaced along the two ledges on either side of the central channel. Based on harmony and resonance, I proposed that frames were inserted into the slots and held in place vertically using the grooves that were cut into the third corbel.

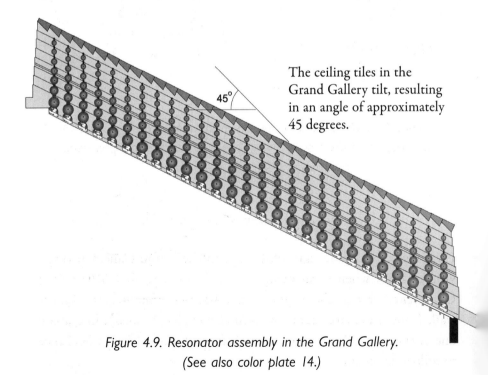

45°

The ceiling tiles in the Grand Gallery tilt, resulting in an angle of approximately 45 degrees.

Figure 4.9. Resonator assembly in the Grand Gallery.
(See also color plate 14.)

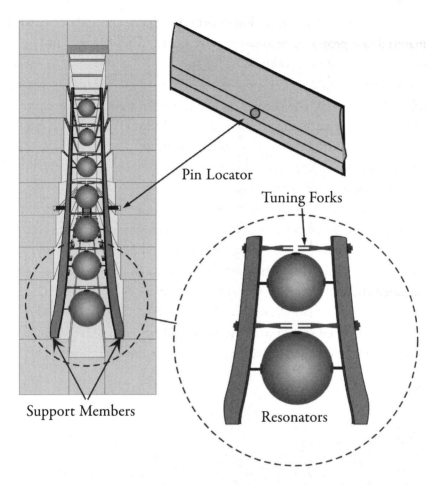

Figure 4.10. Once the lower part is in the ramp slots and the pins are located in the groove, the resonator assemblies are prevented from moving. (See also color plate 14.)

EXPLOSIVE NEW EVIDENCE

I was able to see more signs of damage in the Grand Gallery during a visit in 1999, when I was being filmed during the New Millennium Conference. When the bright camera lights illuminated the gallery wall, I noted damage marks on each corbel that formed a line above one of the slots. As my time was spoken for, I did not have the chance to follow up on this.

Figure 4.11. Photograph of ceiling of Grand Gallery in 2001, taken from the step at the south end. (See also color plate 14.)

Another opportunity came in 2001, when, as I discussed earlier, I was being filmed by Grizzly Adams Productions for a Pax Television documentary that also featured Zahi Hawass. It was a fortuitous time because the pyramid had been closed for cleaning for several months. Rumors spreading around the internet claimed that the reason given for closing the Great Pyramid for cleaning was an excuse to surreptitiously dig a tunnel to connect with the Queen's Chamber South Shaft and access what was behind Gantenbrink's door. Therefore, I was greatly appreciative of Hawass giving Gail Fallen and me permission to enter the Great Pyramid, which was still closed.

Figure 4.12. Resonator frame with scorch marks. (See also color plate 14.)

Little did I know at the time that this visit inside the Great Pyramid would reveal additional critical evidence that was not known when *The Giza Power Plant* was published.

Indeed, there is no question that the inside of the Great Pyramid had been cleaned. Comparing previous photographs of the Grand Gallery with those taken after the cleaning leaves no doubt about this fact. Strikingly, the ceiling of the gallery immediately got my attention and caused my jaw to drop. The tiles above the pairs of slots on each side of the gallery had scorch marks. Evidence supporting the scenario I discussed in 1998 had been revealed after the dust and soot were removed by cleaning. Intense heat created by burning resonators had left an indelible signature etched into the limestone tiles.

THE AFTERMATH

There is no shortage of suppositions and probabilities about what happened after the Great Pyramid was violently shaken and shuddered to eerie stillness. How extensive was the wreckage inside? Early explorers were faced with clearing tons of material from the Descending Passage, the Subterranean Chamber and Well Shaft, and the Pit that begins at the bottom of the Grand Gallery and connects with the Descending Passage just north of the entrance to the Subterranean Chamber. Masonry with unusual shapes and features was noted, such as half drill holes evident along the edge of a piece of granite. Among the debris was a prism-shaped piece of stone, which was surmised to be the piece that covered the granite plugs located where the Descending and Ascending Passage meet. That supposition may be correct, although another possible location for the piece may have been in the Grand Gallery itself. Was it one of many that were positioned on the ledges on both sides of the Gallery so as to match the tiled ceiling? It's a point to keep in mind when considering restoring the pyramid to its functioning status.

Reports of debris being cleared away to gain access to various parts of the pyramid leave a question mark regarding what was taken out and what its function was in this electron harvesting machine. A compelling story has taken firm hold of the imaginations of scholars, researchers, and readers about the history of discovery within the Great Pyramid. Specifically, the adventures undertaken by Caliph al-Ma'mun who gained entry into the Great Pyramid in AD 812 with great effort by carving a tunnel from the outside to the Descending Passage. Reportedly, he had searched the north face of the pyramid for days looking for the entrance, and after reaching frustration with his lack of success, began lighting fires and quenching the hot stone with vinegar to persuade the limestone to yield to the action of his hammers and chisels more easily.

His mission was inspired by rumors of untold treasures held inside, such as glass that might bend but will not break. Recovering these treasures seemed to overshadow more unseemly rumors about naked women with large teeth who waited to seduce young men. But, on the other hand, perhaps those rumors were an attraction also?

Nevertheless, it is reported that Mamun and his men chiseled away for several months before breaking through into the Descending Passage. However, this story is questioned by researchers Ralph Ellis and Mark Foster in their article "The Enigma of Mamun's Tunnel," in which they present a cogent argument suggesting that rather than cutting the tunnel from the outside in, Mamun cut it from the inside out after gaining access to the inside of the pyramid the same way Strabo did in 24 BC—through the original entrance.

> The question is, therefore, why could Mamun not see these tell-tale marks and the original entrance to the pyramid, that lay only a few meters above him? Why could he not see the handle on the door, or the scuff-marks on the smooth exterior? And knowledge of the location of the true entrance must still have been known in this era, so why could none of the locals be 'persuaded' to point it out? And this apparent invisibility of the original entrance could not have been because it was covered by sand, for instance, because Mamun's tunnel lies below the level of the real entrance. So what was the problem? Why was so much effort expended in digging a new tunnel, when the original entrance lay just above it? (Ellis and Foster 2000, 42)

Ellis and Foster answer this question by suggesting that Mamun discovered and removed more from inside the Great Pyramid than has officially been revealed:

> It is highly probable that the real reason for constructing the forced tunnel was not to get into the pyramid, but rather to get something out. Whatever it was, though, it must have been small enough to go down the first part of the ascending passage, but it was too long to go around the bend between the descending and ascending passageways. The only alternative for these intrepid but highly destructive explorers, was to dig a tunnel directly outwards from the

junction of the two passageways, completely bypassing the internal passageway constriction. (Ellis and Foster 2000, 42)

Ellis and Foster proposed that it might have been the lid from the sarcophagus that could not be wrestled around the bend to be pushed up the Descending Passage and through the original entrance. Knowing the weight of similar items that have been found, one would surmise that such an item would have been extremely heavy, and lowering it down and through a horizontal tunnel was preferred to pushing it up a passageway angled at 26 degrees. The other possibility is that there were items that were recoverable and able to navigate the Ascending Passage but too long to make the turn at the Ascending/Descending Passages junction.

I fully support Ellis and Foster's analysis. Their arguments against Mamun's generally accepted reason for digging his tunnel and the alternative scenario they present make perfect sense. The question is, other than a sarcophagus lid, what else might Al Mamun and his men have carried out of the Great Pyramid that was so valuable that they went to the trouble of digging their tunnel?

Immediately, my thoughts turned to the question of what happened to any possible remnants of the Grand Gallery resonators. If some of the tall frames survived the fire, most certainly they could not have been taken up the Descending Passage. It boggles the mind to think that perhaps somewhere in Egypt a stash of materials exists that may once have been housed in the Grand Gallery and performed in harmony with the Earth inside a mountain of stone. However, nothing has officially been disclosed about any stash of materials related to vibration, frequency, harmony, and music that could be interpreted to have come from the Great Pyramid. Unofficially though, in 1997 a tantalizing post appeared on a Keelynet bulletin board that disclosed a clandestine and unauthorized visit to a mysterious room in an Egyptian museum.

"Anomalous" Ancient Egyptian artifacts
    Author: No name given.
    Date: Wed, 10 Dec 1997 17:10:09 +0000

Some years ago an American friend picked the lock of a door leading to an Egyptian museum store-room measuring approx 8 feet × ten feet [sic]. Inside she found "hundreds" of what she described as "tuning forks".

These ranged in size from approx 8 inches (20.03 cm) to approx 8 or 9 feet (2.44–2.74 m) overall length, and resembled catapults, but with a taut wire stretched between the tines of the "fork". She insists, incidentally, that these were definitely not non-ferrous, but "steel".

These objects resembled a letter "U" with a handle (a bit like a pitchfork) and, when the wire was plucked, they vibrated for a prolonged period.

I know of nothing that can corroborate the story above, but I include it out of interest because when I created the illustration of the resonator for *The Giza Power Plant*, I included tuning forks as part of the assembly. I also included what I described as Helmholtz resonators, which worked in unison with the tuning forks. While the story should be treated with some measure of doubt until strong evidence is revealed that would support it, I can't help wondering why someone would make the story up. I can say with certainty that if it is a fabrication, *The Giza Power Plant* did not inspire it because the post appeared eight months before my book was published. Neither did this story inspire what is written in my book. When the post appeared on the bulletin board, my manuscript was already finished and at the publisher, and I did not learn about it for a long time afterwards. Perhaps more information will be revealed someday that will corroborate the story, but for now it will remain a tantalizing mystery that finds a home among many other tantalizing mysteries.

RESONATOR CLARIFICATION

The theory regarding Helmholtz resonators being installed in the Grand Gallery is not without challenge. But then, everything is subject to challenge, and rightly so. Carr shared his opinion that the resonators inside the Grand Gallery may have functioned less like a Helmholtz resonator, such as the way that the hollow body of a stringed instrument like a guitar performs, and more like the strings on that instrument. (An assembly of large, strung tuning forks, perhaps?)

Carr's opinion was based on his analysis that the King's Chamber, by its geometry and dimensions, functioned as a large Helmholtz resonator itself. Carr's professional analysis, when combined with Robert Vawter's computer modeling, provides support to the observations made by flautist Paul Horn, who played his flute on the Great Step at the top of the Grand Gallery and noted that the gallery response was flat, but from behind him in the King's Chamber he heard a strong reverberation of the sound. These analyses are in agreement with my own observations when in 1995 I played back the tape that was recording inside the King's Chamber when I was in the Grand Gallery calling for the guard to turn off the lights. The playback on my tape recorder was as though I had not left the chamber (Dunn 1998, 161–162).

Carr may be correct, but there is another aspect of ancient acoustics that Robert Vawter recently shared with me that might contribute to the discussion—the historical use in Roman amphitheaters of Helmholtz-type vessels that were placed within cavities beneath the terraced seats to enhance the sound. In Appendix A, Vawter will discuss this in more detail and draw upon Vitruvius's work, *The Ten Books on Architecture* (Vitruvius 1914). See also color plates 15 and 16 for an illustration of acoustic phenomena observed by Vawter in the Grand Gallery.

It would be a mistake, however, to isolate the Grand Gallery and King's Chamber as the only features that were acoustical in nature. The entire pyramid and all its parts demonstrate unique acoustic properties—with reports of sound originating in the Subterranean Chamber being heard in the King's Chamber, the passageways that carry the sound through the pyramid are analogous to organ pipes, as Vawter will discuss in his appendix. Obviously, as anyone who has crawled through these passages will tell you, they were not designed to welcome and provide a comfortable route for humans to reach their destination.

Then there is the Antechamber. I have little to add to what I wrote in *The Giza Power Plant* about this critical element, which, I speculated, served as an acoustic filter. But, again, I am not the final judge of what it was used for or the physics behind it. These answers must come from experts like Carr or others with similar education and training.

The same conclusion applies to my analysis of the Queen's Chamber. While the next two chapters will reveal that my principal observations and predictions have been supported by evidence, there are still better qualified specialists than I whose professional analyses, as you will find in the next chapters, carry more weight.

FIVE

THE QUEEN'S CHAMBER SHAFTS

The world's fascination and focus always turns to Egypt and its mysteries when a new discovery is made. Following a string of explorers and researchers before and during the Victorian era, fascination with Egyptian tombs and what they might reveal reached a fever pitch in 1922 when archaeologist Howard Carter (9 May 1874–2 March 1939), working under the patronage of George Herbert, the 5th Earl of Carnarvon (26 June 1866–5 April 1923), discovered Tutankhamun's tomb and revealed a treasure trove of priceless funerary artifacts to the world.

With this discovery, the question of the existence of secret chambers containing Khufu's mummy inside the Great Pyramid gained more relevance. If a tomb with such fabulous riches as Tutankhamun's could exist in a dusty valley on the west bank of the Nile near Luxor, surely it would be quite reasonable to assume that a grand pyramid—the largest structure that had ever been built—would be the repository of treasures far more plentiful and extravagant in design and materials? The apparent wealth available to the people who invested in such a stupendous structure would surely dwarf every artifact that has been found in every tomb discovered in the Valley of the Kings and elsewhere in Egypt.

If we look back at the numerous attempts to discover something new inside the Great Pyramid, we can identify a single motivation for those who held the keys to this edifice: the search for the undisturbed

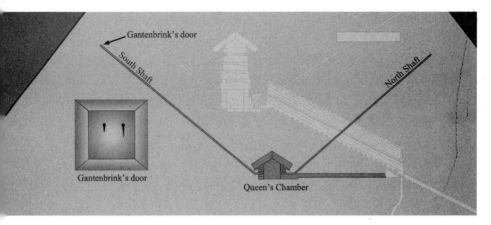

Figure 5.1. The Queen's Chamber shafts and Gantenbrink's door.

tomb of Khufu, or Cheops as he is also known. Although the three chambers we know about have yielded nothing, Egyptologists still believe that evidence of the king's interment can be found.

I had studied the Queen's Chamber and its mysterious and confounding shafts for a long time, trying to figure out how they may have functioned in what I had come to believe was a massive machine. I accessed through the library or purchased just about every book I could find that might have some information about the shafts. One of the most useful reports of their discovery can be found in Charles Piazzi Smyth's book, *The Great Pyramid: Its Secrets and Mysteries Revealed.*

Newly discovered Air Channels in the Queen's Chamber.

Now in what is just passed, we have seen a whole series of connections between the actually existing measurable facts of the Queen's Chamber, and scientific portions of the ultimate, and originally secret, design of the Great Pyramid. Therefore, although some of the early travellers [sic] have spoken fearfully of "the grave-like smell and noisome odour [sic] of this room, causing them to beat a rapid retreat," the room must have acquired that revolting character from modern vilifying, rather than ancient construction; for what its builders put into it, as we see above, is not of a nature to experience any fleshly corruption.

Indeed, in its ancient planning, the Queen's Chamber would appear to have been intended someday to be remarkably well ventilated. For the chief item of latest discovery at the Great Pyramid, is that one which was made in 1872 by Mr. Waynman Dixon, in company with his friend Dr. Grant, and with the assistance of one of his English workmen from the bridge he was then erecting over the Nile; and is to the effect that this Queen's Chamber has, in a peculiar state of readiness, though never yet brought into action, two ventilating channels in its north and south walls, nearly similar to those in the King's Chamber.

Perceiving a crack (first, I am told, pointed out by Dr. Grant) in the south wall of the Queen's Chamber, which allowed him at one place to push in a wire to a most unconscionable length, Mr. W. Dixon set his carpenter man-of-all-work, by name Bill Grundy, to jump a hole with a hammer and steel chisel at that place. So to work the faithful fellow went, and with a will which soon began to make a way into the soft stone, when lo! After a comparatively very few strokes, flop went the chisel right through into somewhere or other. So all the party broke away the stone round about the chisel hole, and then found a rectangular, horizontal, tubular channel, about 9 by 8 inches (22.86–20.32 cm) in transverse breadth and height, going back 7 feet (2.13 m) into the wall and then rising at an angle of about 32° to an unknown dark distance.

Next, measuring off a similar position on the north wall, Mr. Dixon set the invaluable Bill Grundy to work there again with his hammer and steel chisel; and again, after a very little labour, flop went the said chisel through, into somewhere; which somewhere was presently found to be a horizontal pipe or channel of transverse proportions like the other, and, at a distance within the masonry of 7 feet, rising at a similar angle, but in an opposite direction, and trending indefinitely far.

Fires were then made inside the tubes of channels; but although at the southern one the smoke went away, its exit was not discoverable on the outside of the Pyramid. Something else, however, was discovered inside the channel, viz. a little bronze grapnel hook; a portion of cedar-like wood, which might have been its handle; and a grey-granite, or green-stone ball, which from its weight, 8,325 grains, as weighed by me in November, 1872, must evidently have been one of the profane Egyptian mina weight balls, long since valued by Sir Gardner Wilkinson at 8,304 grains.

To this passage, Smyth added the following footnote:

Thin flakes of a very white mortar, exuded through the joints of the channel, were also found; and being recently analyzed for me by Dr. William Wallace of Glasgow, were proved to be composed not of carbonate, as generally used in Europe for mason's mortar, but sulphate, of lime; or what is popularly known as "plaster of Paris" in this country. (Smyth 1978, 427–429)

There is a full chapter in *The Giza Power Plant* that discusses the history of exploration and discovery in the Queen's Chamber and its shafts before 1998. Within the chapter, I presented a formula given to me by chemical engineer Joseph Drejewski for creating hydrogen. His formula specified the mixing of two chemicals, hydrated zinc solution and a dilute mixture of hydrochloric acid and water, with the hydrochloric acid feeding into the chamber through the South Shaft and hydrated zinc through the North Shaft. Obtaining these chemicals, I mixed them in a vessel with a "chimney" on top and was able to demonstrate its functionality by igniting the hydrogen gas that was created from the reaction.

I also made an attempt to explain Dixon's discovery of a grapnel hook, cedar-like wood, and stone ball by theorizing about their function within the shaft. However, for the purpose I proposed, which was to create an electrical circuit, they are not necessary, so I must take them out of the equation and place a question mark as to why and for what purpose they were there. A logical reason would be they were used in an attempt to explore the shaft and possibly retrieve what may be there.

Recognizing that other formulas may be defined, I also suggested that sulfur could have been included in one, which may explain the

Figure 5.2. Using Drejewski's formula to create hydrogen.

existence of the "grave-like smell and noisome* odor" of the room that caused early explorers to "beat a rapid retreat."

1. Was the disgusting smell that caused early explorers to beat a hasty retreat from the chamber connected to a chemical process that used sulfur? Hydrogen sulfide is particularly odorous, exuding a smell similar to rotten eggs. This gas is formed by the combination of sulfur with hydrogen. While early explorers expected to be confronted with a certain amount of bat dung inside the pyramid, it seemed as though the smell in this particular chamber was more pronounced than in the rest of the pyramid. Again, the composition of the salts on the chamber's walls may help clarify this investigation, as sulfur-bearing compounds may have formed these salts.

2. Where did Caviglia get the chunks of sulfur that he burned in the Well Shaft? While it was a practice of early explorers to burn sulfur to purify unhealthy air, it would be most helpful to know whether or not he had the sulfur with him or if it was already there.

---

*"Noisome" means offensive to the point of arousing disgust; foul.

3. If a chemical exchange process was separating hydrogen and if a catalyst was being used in the Queen's Chamber, would sulfur play a part in the operation or perhaps regeneration of the catalyst? (Dunn 1998, 195–196)

The questions asked were, as it turns out, reasonable and logical. In 2015, Brett Cohen, PhD, published an article in *Nexus* magazine that examined my theory regarding the shafts and proposed an alternative chemical process that included sulfur as one of the possible elements used. If Cohen's formula was used by the ancient pyramid builders, it may explain Smyth's anecdotal report of early explorers "beating a hasty retreat." His article is reprinted with permission as Appendix D in this book (Cohen 2015, 43–46).

I know Drejewski's formula works, and I demonstrated it publicly for the History Channel's *Ancient Aliens* show. I accept Cohen's formula and thank him for his analysis. The reason Cohen's formula seems to fit more accurately is that it explains the horrible smell suffered by early visitors to that chamber. But how could this smell have survived the passage of time since the pyramid was first built and used? The other question is, do other pyramids, assuming that they were built for similar purposes, exhibit the same lingering odor of ancient chemicals?

Well, at least one does that I have experienced. A single chamber in the Red Pyramid at Dahshur smells strongly of ammonia. Why? Was Cohen's other chemical solution employed in this pyramid? This is the uppermost chamber in the pyramid, which can be reached by passing through other chambers and climbing several flights of recently constructed wooden stairs. I obviously have not been in all the pyramids in Egypt, but a lingering chemical smell in a remote chamber would be on my list of observations should I find myself out in the field again.

Until recently, with the muon scanning of the Great Pyramid, and the thermal imaging that revealed anomalies on the east side of the pyramid, attention has been focused for the past 27 years on the Queen's Chamber and its mysterious shafts. Even though I believe that more exciting data could be gathered by a thorough scientific examination—with an emphasis on acoustics—of the King's Chamber and all its parts, including the multiple layers of monolithic granite beams above it, worldwide attention was riveted on mysteries surrounding the enigmatic Queen's Chamber shafts that reach for the sky with no known termination point.

GANTENBRINK'S PYRAMID EXPLORATION

For 121 years following Waynman Dixon's discovery, the dark mysteries inside the Queen's Chamber shafts tantalized those who pondered on the shafts' reach. Where do they end? Are there secret doors on the outside of the pyramid that can be opened to access their exit point? Then in 1993, a flurry of excitement rippled around the world when information was released about a remarkable discovery inside the Great Pyramid. German robotics engineer Rudolf Gantenbrink, who had been contracted by the German Archaeological Institute in Cairo to install ventilation fans inside the pyramid's King's Chamber shafts, requested and received permission to design a robot that would explore the shafts inside the Queen's Chamber to try and answer the more-than-century-old question. Unlike the shafts leading from the King's Chamber, these approximately 8 inch (20.32 cm) square shafts had no known exit point on the outside of the pyramid.

The robot, named Upuaut II (meaning opener of the ways), was equipped with lights, camera, and laser pointer, and has received more fame and recognition that any other similarly equipped explorer on the subject. What the robot revealed to the world was a termination point high in the body of the pyramid. After traveling 220 feet (67 m) at an angle of 39 degrees, Upuaut II came to a dead end when it was faced with a solid limestone block barring the way. Protruding through the block were two mysterious, presumed-to-be-copper pins that were bent down against the surface of the rock. In *The Giza Power Plant*, I described these copper pins as electrodes that maintained head pressure by signaling the need for more chemicals to be pumped into the shaft when the liquid dropped. I theorized that the chemicals were pumped up a vertical shaft that connected with a space behind Gantenbrink's door.

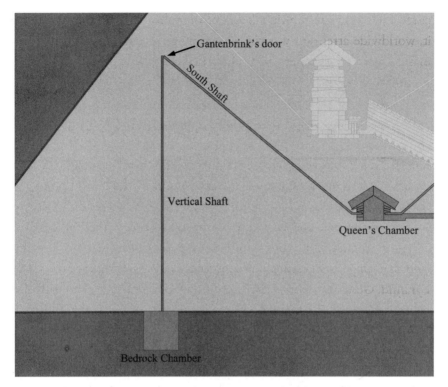

Figure 5.3. Queen's Chamber South Shaft with vertical shaft.

The discovery of this barrier posed a dilemma for Egyptologists. Simply by virtue of its existence, the popular belief that the King had changed his mind about being entombed in the Queen's Chamber and decided to have his eternal resting place higher up in the pyramid was brought into question, because the termination point of the shaft was higher than the King's Chamber. The obvious question arising from this discovery is, "Why continue to construct these shafts that did not connect with the Queen's Chamber or the outside of the pyramid if the Queen's Chamber was not going to be used?" When Gantenbrink sent his robot trekking off into the unknown, Egyptologists did not know that what the robot would discover would challenge their theory and call it into question. They had to provide an answer for the discovery, and one that would make sense. As it was established that deceased kings were not buried with their queens, they modified their theory and now claim that the Queen's Chamber was a dummy chamber and that the shafts were symbolic features through which the King's soul exited the pyramid, by magical means, into the sky. Gantenbrink, quite properly, considering the shaft is a mere 8 in (20 cm) square, dubbed the blocking stone a USO (Unidentified Stone Object). However, the official description was "door," and the popular reference to the discovery became known as Gantenbrink's door. Was it because a door inspires the possibility of something existing behind it that Gantenbrink's preferred nomenclature was ignored?

For the theory proposed in *The Giza Power Plant* to have validity, whatever is discovered behind Gantenbrink's door must be reconcilable within the context of the pyramid being a power plant—or electron harvester. Adhering to scientific principles, what I had theorized should be falsifiable. In other words, it should be available and subject to scrutiny, testing, and verification to determine without doubt whether it is either true or false. If a royal chamber full of funerary trappings and a mummy is discovered beyond this barrier, then what I have proposed in my book is severely weakened and may be null and void. Within the framework established by Egyptologists, based on the standards they have established for the explanation of evidence with the pyramids, whatever is discovered can be explained away by invoking mysterious and magical occult symbolism, a theory that is impossible to prove and is, therefore, unfalsifiable and unscientific.

There is no mystery about where the King's Chamber shafts end. They exit the pyramid cleanly with no ambiguity. The Queen's Chamber shafts, however, do not, and the Egyptologists' hope was that they lead to the long-wished-for, sought-after, treasure-laden burial room of King Khufu. This hope has been, and still is, the impetus for all official explorations in the Great Pyramid.

The tools and methods employed over the past couple of centuries have changed from the aggressive ripping out of pyramid stone by setting off dynamite and attacking the limestone with pickaxes and chisels, as famously performed by Howard-Vyse in 1837, to the more recent, sophisticated muography methods employed by the ScanPyramids mission, and everything in between. When these explorations reached their conclusion and the tools were packed and shipped back to wherever they originated, the search for burial chambers has always come up empty. No burial chamber. No Khufu's mummy. What has been exposed, however, has been more evidence of the critical functions of this sophisticated precision electron harvesting machine.

THE PYRAMID ROVER EXPLORATION

After *The Giza Power Plant* was released, I, like millions of other Egypt watchers around the world, was waiting for the day when additional explorations would take place and another tantalizing barrier to greater knowledge might be removed. Any news that came from the Giza Plateau I immediately examined and researched. Some information I commented on at the Giza Power website and I also published articles in *Atlantis Rising* magazine. Around that time, there was an inhabitant of Giza who reported on happenings around the plateau. He even set up a camera that was focused on the pyramids and ran a live feed to the internet. With such huge interest around the world and a fledgling online social media capability growing quickly, a live stream of little going on was considered entertainment.

On September 16, 2002, hundreds of millions of people around the world tuned in to watch "Opening of the Secret Chamber: Live!" The FOX network in the US broadcast to 30 million viewers, and in China alone, it was reported that the event attracted half a billion viewers. It had been nine years since Rudolf Gantenbrink made his famous discovery of the limestone barrier at the end of the Queen's Chamber South Shaft. Had Gantenbrink been allowed to continue his explorations after his 1993 discovery and equip his robot to penetrate his "door," as he wanted to, we would have known what was behind it a long time ago. But now, a new robot—aptly named Pyramid Rover—was in town. Built by iRobot Corporation in Bedford, Massachusetts at a cost of $250,000, its stated mission was to drill a hole in Gantenbrink's door, insert a camera, and see what lay beyond.

At the time *The Giza Power Plant* was published, this exploration was not on the horizon, and I had no idea when more information would be

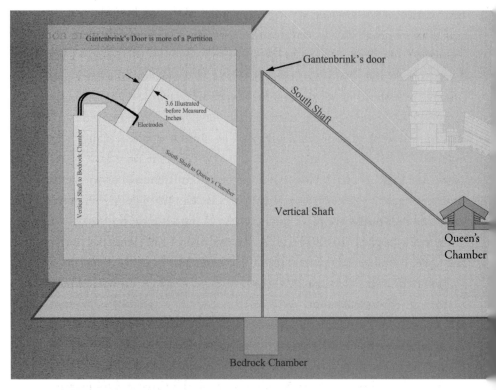

Figure 5.4. Original prediction of what may be found behind
Gantenbrink's door.

revealed, but, anticipating that further explorations would be made someday, I included in the book a prediction of what might be found when the
day came. This prediction was based on what is found elsewhere in the
Great Pyramid and is an engineering analysis of the function of the pyramid as a machine. The prediction was that behind Gantenbrink's door
would be a vertical shaft that connected with a bedrock chamber beneath
the Great Pyramid. The bedrock chamber, I proposed, would house the
mechanism necessary to pump chemicals through this supply shaft to the
Queen's Chamber shafts to sustain the production of hydrogen for generating maser activity in the King's Chamber. See color plate 17.

When I visited Egypt in 2001, after the book was published, I
became aware of a feature to the east of the Great Pyramid that seemed
to suggest a much better design was used than a vertical shaft. A curi-

ous vertical shaft outside the pyramid near the center of the north-south axis caused me to consider that perhaps the chemicals were not pumped from a bedrock chamber under the base of the pyramid but from outside the pyramid, through chemical supply shafts that I proposed were connected to the Queen's Chamber South and North shafts. Logistically, it makes more sense not to have to go into or under the pyramid to replenish the chemicals. The supply shafts leading from external chemical containers up to the Queen's Chamber South and North shafts could have been included in the construction as the layers of the pyramid were built. Interestingly, this change in design is not the only one that makes more sense than what I proposed in my book.

On September 16, 2002, the exploration of the South Shaft in the Queen's Chamber of the Great Pyramid was broadcast live from Egypt. Before the show aired, a mixture of predictions on what would be found behind the "door" were discussed in various forums, websites, and broadcasts. Most of them were clearly based on the Great Pyramid being the tomb for King Khufu. They included:

- A sacred book written by Khufu would be discovered—perhaps Khufu's diary or guidance on navigating the afterlife.
- A sacred scroll.
- A statue that was positioned to view the stars.
- It doesn't matter what they find. Unless it's an alien spaceship, it will support the tomb theory.
- A space 30 feet (9.14 m) long that contains sacred sand.
- Nothing. Just core masonry.

On September 15, 2002, I openly shared the prediction I had published in my book, with an update regarding the external storage and pumping station, on my website and in radio interviews about what might be found. When George Noory on *Coast to Coast AM* asked me what would become of my prediction if behind the "door" they found scrolls, a statue, or a burial chamber, I responded that this would most likely support the tomb theory and essentially put to rest any arguments to the contrary—including mine.

Figure 5.5. Vertical shaft on east side of Great Pyramid, looking north.

Figure 5.6. Vertical shaft on east side of Great Pyramid, looking west.

Figure 5.7. Vertical shaft on east side of Great Pyramid, looking east.

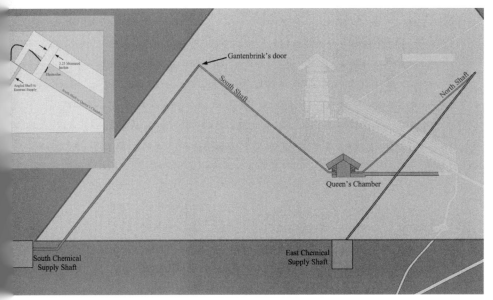

Figure 5.8. 2002 revision of theory to include external shafts.

The two-hour FOX/National Geographic extravaganza provided a breathless prelude to the moment when Pyramid Rover's masonry drill bit finally broke through into the space beyond and the endoscopic camera was inserted into the hole to take a peek at what lay beyond Gantenbrink's door.

The preparatory work was fascinating. With necessary due diligence, the team completed an exhaustive research and experimental period, which included:

- Measuring the thickness of the stone blockage using an impact sensor, which uses reflected sound waves to determine distance. (Before this test was conducted, the right metal pin was shortened to allow the sensor to fully connect with the limestone.)
- Building a replica of the South Shaft on the same angle, including a limestone block with the same dimensions as the shaft and the recently measured thickness.
- Equipping the robot with the correct drill size to drill a hole in the test "door" through which the robot could insert a camera to capture on video what was beyond.
- Hawass and his colleagues observed the drilling of the rock and gave approval for the team to proceed and drill through Gantenbrink's door.

The illustration in my book closely predicted one of the attributes of the "door," and the evidence supported this prediction. In my illustration, the thickness of the block is given, by scale, as 3.64 inches (9.25 cm). My measure was based on nothing more than my proposed function for the block. The impact sensor on the Pyramid Rover measured the actual thickness and found it to be 3.25 inches (8.25 cm) thick. Function = design = measure. And in reverse engineering: measure = design = function.

Hawass's confidence seemed to become slightly muted as the program drew to its climax. He cautioned the viewers that there may be nothing behind the door at all. His prophetic comments became a reality to all of us as the camera with its fish-eye lens pushed through the hole and a distorted image came into view. There appeared to be nothing there but a rough-looking block a short distance away.

After two teasing hours leading up to the actual penetration of the "door" in the last five minutes, images from the camera feed allowed Zahi Hawass, then chairman of the Supreme Council of Antiquities (SCA), to dramatically announce, "It's another door! With a crack!" Not much was said after that except Hawass claiming that it was a very important discovery. In a later interview he advised that "everything now needs a careful look. We will ask the National Geographic Society to cooperate to reveal more mysteries."

ANOTHER "DOOR?"

On September 23, 2002, news came out of Egypt that the Pyramid Rover team had successfully explored the Queen's Chamber North Shaft in the Great Pyramid. This shaft, directly opposite the South Shaft, posed problems for early explorers, who pushed rods into it in an attempt to explore its depth. In 1993, with Upuaut II, Gantenbrink was unable to navigate around earlier explorers' rods that were jammed in the passage as they attempted to push the rods around a bend in the shaft. The iRobot team had a cunning but simple solution to the problem. They turned the robot 90 degrees and sent it up the shaft gripping the walls instead of the ceiling and floor. In this manner, it was able to ride over the top of the obstacles.

The journey the Pyramid Rover took through the shaft was not advertised nor were any details released, except for what was revealed when it reached its destination—another barrier that was similar to Gantenbrink's door, with two metal fittings protruding through it. There are some remarkable differences between the North Shaft metal fittings and the South Shaft metal fittings, though it was still awarded the same description of being a "door," with the exception of not having Gantenbrink's name attached to it.

Within the context of the power plant theory, the existence of these metal fittings is predictable. Also predictable are the characteristics of the fittings. The power plant theory describes a chemical process where two solutions combine to create hydrogen. I had proposed that the South Shaft provided a diluted hydrochloric acid solution to the Queen's Chamber and the North Shaft provided a hydrated zinc solution. The function of these metal fittings was to serve as anode and cathode electrodes, where the flow of electricity from the anode through the liquid to the cathode would create a circuit and allow monitoring of the fluid levels in the shafts. In the Queen's Chamber North Shaft, however, the liquid contains metallic atoms, which would cause electrodeposition on the cathode.

DOESN'T HOLD WATER

There have been objections to the theory that chemicals were being fed through these shafts into the Queen's Chamber. However, I believe that there is more evidence to support the idea than there is to dismiss it. One of the objections, based on what is observed inside the channel, is that some blocks, having shifted, would cause leakage. I would respond that the channels were constructed to pass fluid, not hold fluid. Similar to a storm sewer, leaks may be present, but the bulk of the stormwater is passed through. Also, the seals on storm sewers cannot be seen on the inside, so without knowing how the shaft blocks are fitted, it is not a valid argument.

What kinds of seals may have been used by the ancients? Evidence of plaster of paris has already been noted by Piazzi Smyth. It was also noticed by remote sensing archaeologist Meg Watters. During the filming of the Pyramid Rover's exploration of the South Shaft, in "Into the Great Pyramid" at timecode 45:05 (National Geographic 2002), Watters noted that mortar appeared to be oozing from underneath Gantenbrink's door. Both Smyth and Watters assumed at first glance that the white substance between the joints of the Queen's Chamber shaft blocks was mortar. Smyth was able to retrieve a sample near the opening of the shaft, and it was determined to be plaster of paris. One can assume that Watters, not being able to retrieve a sample and have it tested, was observing plaster of paris as well. It could be that it was not used to provide mechanical strength to the joint, but to seal the joints and prevent leaks. To my knowledge, nowhere else in the Great

Pyramid has plaster of paris been found oozing from pyramid block joints except the shafts of the Queen's Chamber.

Also, the channel block above a recess in the floor of the shaft, which Gantenbrink described as the tank trap, appears to be sitting on a material that is oozing into the space (See "Into the Great Pyramid," timecode 43:00) (National Geographic 2002).

Another objection to my theory is that using hydrochloric acid solution would corrode the limestone. In this case, the reason for opposing the theory is the same evidence I used to support it. The evidence supporting the theory of dilute hydrochloric acid can be seen as erosion on the walls and floor of the shaft. This corrosion obviously occurred after the pyramid was built. The theory proposed in *The Giza Power Plant* also points out that the metal fittings known to exist in the South Shaft, being electrodes that served to signal when the level of chemicals in the shaft dropped below a certain level, had been corroded also. The left electrode, significantly, had corroded so much that a section of it had broken off. Without any means of causing this corrosion, except for a patina that had developed over time, the metal would be just like it was after installation. After examining all objections, it is my contention that the evidence present supports the theory and does not weaken it. There is ample evidence to prove that the North and South Shafts in the Queen's Chamber display characteristics that indicate some action took place in them after they were constructed. Considering that they were intentionally designed and constructed to be closed on both ends and inaccessible from the chamber they are almost connected to, all the evidence points to a fluid being the source of the action. The open question is what chemicals was the fluid carrying and what was the ratio between the chemicals and water?

Hawass had predicted that behind the stone block at the end of the North Shaft would be "another door." With the exception of the nomenclature of "door," on this point Hawass and I agreed. If and when an expedition is mounted to drill through the North Shaft block, most likely another space similar to the one at the end of the South Shaft will be revealed, as well as a chemical inlet shaft on the east side of the cavity, perhaps in the floor but possibly in the east wall. The reason I said this is because of the external vertical shafts on the east side of the Great Pyramid. See the images on pages 134 and 135.

In the power plant theory, every architectural element in the Great Pyramid is integrally linked. Some features can be analyzed separately, but for the most part, the Subterranean Chamber, Queen's Chamber, King's Chamber, and Grand Gallery are the principal features that work together in unison, and they cannot be separated from each other when considering a piece of evidence. Top to bottom, they depend on each other.

The features found in the King's Chamber, mainly the width and height of the North Shaft, led me to propose the use of dilute hydrochloric acid in the South Shaft and hydrated zinc in the North Shaft of the Queen's Chamber. The features in the Grand Gallery led me to understand the function of the King's Chamber. The features in the Queen's Chamber indicate that a chemical reaction could have taken place there to create hydrogen. The hypothesis rises or falls on the evidence found in these areas. For the theory to hold together, evidence discovered in the future has to support it. Some evidence, such as what will be found behind Gantenbrink's door, can be predicted by what is found in the chamber, the South Shaft, and the North Shaft. The power plant theory will be either vindicated or severely challenged or even dismissed based on what is revealed.

My confidence in the validity of the theory may have caused me to jump the gun after the Pyramid Rover expedition and misinterpret what the images revealed. Like everybody in the US, I was watching the video on FOX television. In the top left corner was the word LIVE and the bottom left carried the FOX symbol with Channel 27. There was really nothing for me to become excited about on this broadcast until a man in Germany provided a high-resolution image he had taken of the National Geographic program being broadcast on Sky television in Europe. This image seemed to indicate there was more to be seen in the area occluded by the FOX logo.

I copied the image into a graphics program and auto-adjusted the levels, which lightened the dark areas. I stared for what seemed to be eternity at what was revealed.

I know that if you stare at something long enough you might be able to see a face or some other shape, but the rectangular shape in the left corner of the new block became immediately apparent. I then adjusted the levels, curves, and colors to bring more definition to the image and created construction lines using the bottom corners as guides to create a vanishing point. It was my intention to see if the geometry of the rectangular shape on the left side was indeed a true rectangle parallel with the wall.

I saw what I was hoping to see. A small, dark, rectangular shadow on the floor next to the east wall and the newly discovered "door." I thought the inlet shaft had been revealed, and I published updated drawings incorporating this new "discovery" and promoted the new information on the radio and on my website.

Unfortunately, I was wrong. But it was going to take 10 years before the evidence revealed by the Djedi robot expedition made me realize it. This remarkable exploration will be discussed in the next chapter.

On September 23, 2002, an additional piece of critical evidence was recorded by the Pyramid Rover robot. Unfortunately, this evidence received no attention for almost two decades. Since I began writing this book, numerous events that I can only describe as moments of synchronicity have transpired. The events were beyond my contrivance or control but were critically important to how this book developed and the information it contains. Such an event occurred on July 29, 2021, when I received an email from one of my readers in the UK named Ben Fitzpatrick regarding recent revelations discovered in the North Shaft of the Queen's Chamber of the Great Pyramid. Ben writes:

Hi Christopher,

I hope you can still be reached here. I've just re-watched Ancient Architect's new video on the Queen's Chamber Northern Air Shaft. "First Look Inside the Great Pyramid Queen's Chamber Northern Shaft"

It contains a revelation, which I don't believe I've seen in your works or diagrams, but which massively supports your Pyramid Power theory, which is that the cathode sits in a little recess in the stone rather than protruding from it. I'm no scientist, but I've just looked up recessed cathodes and it apparently helps to prolong the life of the cathode by extending how long it takes for the build-up to occur. This recess combined with their physical appearance is clearly enough evidence to say these definitely are anode and cathode rather than ceremonial plug handles!

Forgive me if this was something you were already aware of.

Cheers,

Ben

(Ben Fitzpatrick, email to author, July 29, 2021)

This was the first time I had been made aware of this. I knew about the pins and the presumed electrodeposition on the cathode, so after watching the YouTube video, "First Look Inside the Great Pyramid Queen's Chamber Northern Shaft" (Sibson 2021) referenced by Ben, I contacted the owner and content creator, Matt Sibson, and we had a very cordial conversation. As I was in the early stages of writing this book, I was curious about copyright permissions and where he had received his. He told me that he had received images and two articles from an ex-employee of National Geographic and was using them for

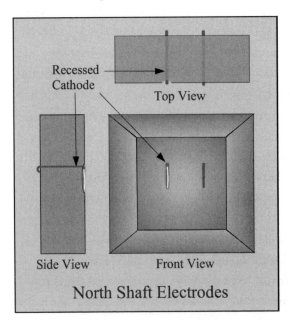

Figure 5.9. North Shaft plug with recessed cathode.

educational purposes, so no release was asked for or obtained. For my purposes, I will not include the photographs here, but will refer the reader to the Ancient Architects video (Sibson 2021) and the relevant timecode of interest (timecode 10:08) and include a drawing of the feature being discussed.

Earlier, I mentioned paying little attention to what appeared in the camera's frame as the robot progressed along the shaft toward its destination. In both the North and South Shafts, the destination was the place where the shaft ended. In the case of the South Shaft, that point was already known, whereas it was still a mystery in the North shaft.

What Sibson had received from his contact were 51 images taken by the Pyramid Rover as it climbed the North Shaft seeking the termination point. This exploration took place in the weeks following the broadcast of the drilling through Gantenbrink's door and was not given the same attention. Nonetheless, what was revealed from this exploration was another blockage with two metal pins similar, but not identical, to those installed in Gantenbrink's door. It was the nature of this blockage and the pins that prompted Ben Fitzpatrick to contact me.

While some of the photographs appeared in a journal article authored by Zahi Hawass that was not widely distributed among the public, Matt noted in his video that there seemed to be a lack of inquisitiveness about the anomalies seen in the photographs.

As I had already commented on the Giza Power website about the appearance of these electrodes in the North Shaft and concluded that the cathode appeared to have been electroplated, Ben's additional observations that a recessed electrode extends its life were just additional support. Interestingly, while most likely not even considering the electrical nature of this discovery, Zahi Hawass had commented on this feature in an article entitled "The So-Called Secret Doors inside Khufu's Pyramid."

At 209 feet (63 m) in the northern shaft (QCN) we discovered a third blocking stone; this closely resembles the first blocking stone in the southern air-shaft (QCS) (Figs 13, 14 and 15). As on the first stone, there were two metal pins protruding from the stone, which are covered with a green patina. The left pin is recessed in a grooved cut in the door and thus sits flush with the surface of the stone. The second pin is bent down on to the surface of the stone itself. The stone appears to be supported by the wall blocks as in the South Shaft. The thickness of this block is not known. (Hawass 2014, 51–68)

In the video, Sibson rightly points out that any theory about the function of the Great Pyramid has to logically be able to explain all evidence found within. I made a similar comment in *The Giza Power Plant*. Also, following the scientific method, if new evidence is found, the proposed theory has to accommodate and explain it. If the theory cannot do that, then the theory cannot stand. New evidence cannot be dismissed or ignored.

There is now no doubt in my mind about the electrical function of the metal pins in the Queen's Chamber shaft. The North Shaft barrier had received extra attention and detail, with a groove being cut into it so that the metal pin at just this location can be recessed from the surface of the stone. The metal pin so fitted is the only pin that has a material deposited on it.

The exploration of the North Shaft and what was discovered at the end were predictable and, to my mind, vindicate the purpose for these shafts as outlined in *The Giza Power Plant*. The image of another "door" with metal fittings and the subtle difference between these fittings and those at the end of the South Shaft support the hypothesis regarding the chemicals used. The electrodes are affected by different chemicals in different ways. In the South Shaft, the action of the dilute hydrochloric acid eroded the copper over time. Because the upper part of the copper was covered with chemical for a shorter period of time than the lower part, as the fluid was always falling, the lower part of the copper was eroded more than the upper part. This resulted in a tapering of the copper and the ultimate failure of the left electrode. In the North Shaft, we see a different effect. Because this shaft contained a hydrated metal, such as hydrated zinc, what we see is an electroplating of the anode. In the photograph taken by Pyramid Rover, there is a white substance on the left electrode only. Unlike the electrodes in the South Shaft, there is no erosion on these electrodes and the pins appear to be less thick than those in the South Shaft. Notice also the stained limestone to the left and on top of the electrode.

LOCATION OF THE CHEMICAL INLET SHAFT

I don't want to misidentify a feature again, as I did after the Pyramid Rover Expedition in 2002, but I have to point out that the visuals and surrounding evidence provided by the Pyramid Rover revealed a surprising location where the chemical inlet shaft connects with the North Shaft. In both the South Shaft and the North Shaft, I had assumed that the chemicals entered the shafts at their upper ends, behind the so-called door. I didn't expect to find that the inlet would be lower down the shaft but, considering that a lower connection would be closer to the supply on the outside, it now makes perfect sense.

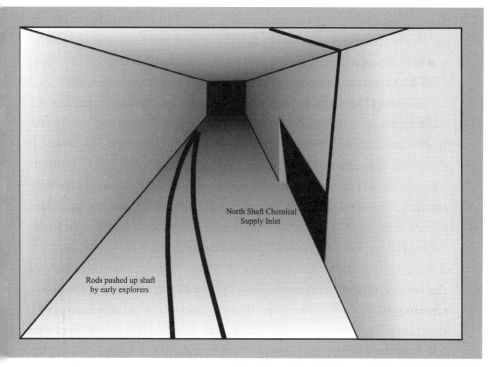

Figure 5.10. Opening in Queen's Chamber North Shaft.

I am now faced with accommodating new evidence and revising my theory again, as the photographs reveal that to all appearances the shaft supplying the chemical to the Queen's Chamber North Shaft is located approximately 96 feet (29.26 m) from the opening of the North Shaft in the Queen's Chamber. I highly recommend that the reader watch the video on the Ancient Architects channel that I referred to above at timecode 10:08, where this feature can be seen on the east wall on the second bend of the shaft.

Opposite the opening there is a collection of trash from where Pyramid Rover retrieved crumpled paper and a site ticket. The site ticket was dated to the 1970s. It is old, but certainly not as old as the pyramid. How did these items get there? It has been speculated that they were introduced into the shaft when the opening was cut after the pyramid was built. My belief is that the opening is part of the original design of the machine and its placement makes it closer to the vertical shafts located outside the Great Pyramid on the east side near the center of the north-south axis, as seen in Figures 5.5, 5.6, and 5.7. Also, as the photograph of this opening shows, the collection of trash is to the left of the rods that were jammed in the tunnel, while the floor in front of the opening appeared to be swept clean. It would seem that a strong wind had carried the trash up the shaft from the outside.

Looking at the photographs of these exterior shafts, we see a vertical shaft cut into the bedrock that has a center wall, creating two shafts approximately three to four feet from the top. If you look down into the shafts, there is trash, blown sand, pop cans, candy wrappers, cigarette butts, pieces of paper, and perhaps a site ticket or two. Also, what would we find if we removed all that trash and sand? How deep does it go? What would we find at the bottom? In 2001, I was shown a vertical shaft, south of the pyramids but a large distance away, that was being cleaned out. At that time, they had reached 100 feet (30.48 m) into the bedrock.

The external shaft seems to be a logical source for the miscellaneous collection of trash. It gathers opposite the opening of the inlet shaft, and some of it slides down the shaft and accumulates. How else would this collection of garbage have got in? The trash could not have come from the upper section of the North Shaft, and it's doubtful that it was carried into the Queen's Chamber and pushed up. The most likely means is that it was blown in by a sandstorm. Maybe an energetic explorer who discovered an opening on the outside started pushing up rods of steel or wood, but I doubt it. It's significant, also, that the floor in front of this opening is seemingly swept clean of debris and sand. This is probably the result of an occasional strong breeze wafting through.

Many questions are left hanging, but the main one at this time is if a connection exists between this external shaft and the North Shaft, where is the opening on the outside? The wall of the shaft closest to the pyramid does not seem to match the bedrock around it, and there appear to be unnatural separations with the adjacent shaft walls, which are clearly cut through bedrock. Was the west wall of the external shaft plastered over at some point?

I had identified this shaft on the outside as the place where chemicals were stored and delivered to the Queen's Chamber shafts. In 2002, I posted an article on my website illustrating this. The newly discovered location of the chemical inlet shaft into the North Shaft makes perfect

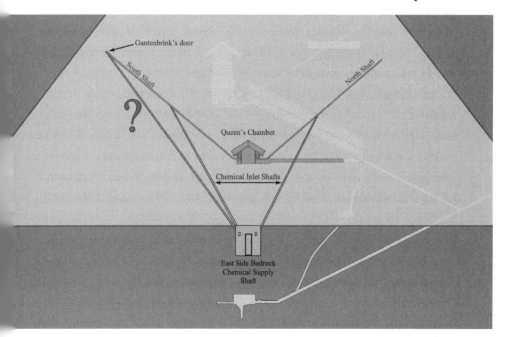

Figure 5.11. Revised 2022 drawing. External bedrock shaft shown greater than scale for clarity.

economic sense, considering that the distance would be shorter than if the inlet shaft was behind the blocking stone. Considering that the North Shaft is approximately 8 inches (20.32 cm) square, the dimensions of the inlet shaft would be about 2.5 inches (6.35 cm).

A useful experiment would be to install in the opening of the North Shaft in the Queen's Chamber a watertight seal that has a hose connector and pump water into it. As the shaft fills, monitor the vertical shaft on the outside to see if it starts showing signs of being saturated or filling with water. Simple test. No robots and very little money. The Queen's Chamber South Shaft still holds the mystery of where the inlet shaft will be found, but it would be useful to conduct the same experiment there too. However, there is an area farther down the shaft from Gantenbrink's door that deserves closer examination.

Following the Pyramid Rover's exploration of the South Shaft and what it revealed about the "door" and what existed beyond, a part of my prediction was proved correct, but other parts were inconclusive. The part that was correct was the function-driven design of the "door," its pins, and the space behind it. The parts of my prediction not proven were the continuation of the pins and any connection they may have had to electrical wiring. Inconclusive also was the existence that I had predicted of a chemical inlet shaft behind the door.

But after what has recently been revealed to a larger public (i.e., the possible existence of an opening lower down in the North Shaft, which I have labeled the chemical inlet shaft), attention should also be given to the possibility that an opening lower down in the South Shaft has previously been overlooked. As I said earlier, with everybody's attention being focused on the destination, much of what exists along the journey has been ignored.

THE PESKY TANK TRAP

Tank traps are used during wartime against enemy tanks. A trench is dug in the ground and camouflaged with brush. The unfortunate tank that travels across this particular section of terrain will find itself crashing through the brush and ending up nose first in the ground, unable to move in any direction. While the builders of the Great Pyramid were not expecting tanks to be moving about in their structure, a feature that is found in the South Shaft was named a tank trap by Rudolf Gantenbrink after his robot Upuaut II found its treads bumped up against a ledge and was unable to navigate further. Gantenbrink's robot found itself in a place in the floor of the shaft that is about two inches (5.08 cm) lower than the rest of the shaft's floor. Not only did Upuaut II have difficulty traversing the tank trap, but in 2002, Pyramid Rover got trapped there also. In both cases, a makeshift bridge was crafted and installed in the shaft so the robots could ride over it. The features of the tank trap can be seen in the video "Into the Great Pyramid," timecode 43:00 (National Geographic 2002).

The only view of the tank trap is from the robot's front camera. A recess in the floor is seen that appears to have been cut intentionally, like a channel through the floor block, except, unlike the channel cut through the block that creates the walls and ceiling of the shaft, this recess is rough cut, and only two inches (5.08 cm) deep. There has not been any discussion about this feature, but it was clearly added to the structure while it was being built and not afterward. I confess to having been puzzled by this recess. At first, my impression was that it was due to erosion and that this stone was, for some reason, more susceptible to the corrosive action of the chemical in the shaft. If that is the case, why just this one block? It also occurred to me that perhaps it was the result of the floor block shifting downwards. But that can't be the answer because the depth of the recess does not reach under the upper channel block but is clearly rough cut, leaving the original upper surface to support the upper block.

Like the South Shaft and North Shaft, the "tank trap" deserves a much closer examination to see if a focus on the destination has missed some critical features along the way.

Perhaps when all the blocks that comprise the South Shaft were installed there was supposed to be a slot cut in the upper block similar to that in the North Shaft. Absent this necessary feature the builders found it more convenient to cut a recess in the floor with an opening in the side that would connect with the chemical supply shaft.

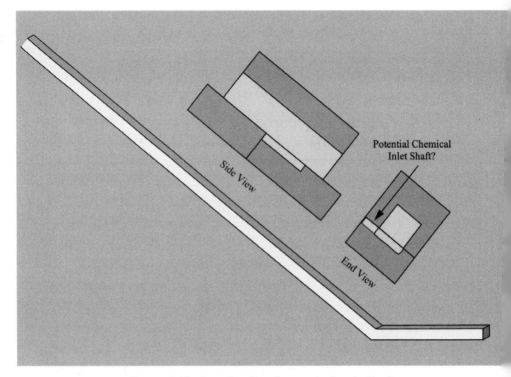

Figure 5.12. Details of tank trap in South Shaft.

Yes, it is speculation, but the fact remains that this feature exists. It is not the result of the bottom floor block shifting downward after it was installed, or caused by erosion. No similar erosion exists elsewhere in the shaft, and the shaft channel block is resting on the bottom block's surfaces that lie outside the width of the recess cut in the bottom block. It was intentional and therefore there must have been a reason for it. Interestingly, the quality of the cutting appears to be identical to the cuts in the North Shaft on the wall near its chemical inlet shaft.

Up until now there has been, as Matt Sibson pointed out about the Queen's Chamber North Shaft, a lack of inquisitiveness regarding what is revealed along both journeys to the prized destination—the infamous "doors." In the next chapter, another expedition explores the Queen's Chamber South Shaft and reveals even more details to puzzle over.

SIX

ENTER THE DJEDI

Be an opener of doors.

RALPH WALDO EMERSON

ANOTHER LOOK BEHIND DOOR NUMBER ONE

It wasn't long after the buzz surrounding the Pyramid Rover expedition inside the Great Pyramid had died down that the world's attention was drawn to the announcement of another expedition that put a different spin on challenges faced by previous robots. This expedition would involve a newly formed team using a different type of robot to explore the Queen's Chamber South Shaft.

As long as the search for Khufu's burial chamber continues, more parts of this energy machine are revealed, which, for me, is a good thing. However, those who were involved in the design and building of the pyramid would probably be quite amused that someone in the future would speculate that behind a blocking stone at the end of a long square shaft that would only allow the passage of a small rodent or robot would be a burial chamber.

THE DJEDI, THE DENTISTS, AND THE PYRAMIDS

There is something about the pyramids that seems to fascinate dentists. Two of my dentists, Dr. Randy Ashton and Dr. Michael Fuesting, have both demonstrated a keen interest in them and the work I have been doing for many years. In fact, Dr. Ashton accompanied me and two other friends to Egypt in 2008.

The Djedi robot expedition was the brainchild of a persistent dentist from Hong Kong, Dr. T. C. Ng, who had made numerous visits to Hawass's office at the SCA to convince him to approve another expedition up the Queen's Chamber shafts using a different type of robot that had more capability than Pyramid Rover. In an interview for a documentary chronicling the expedition, Ng said:

> I went to Egypt with no appointment. From nowhere you come, a Chinese dentist, knocking on the door of Dr. Hawass and ask him to give me the opportunity soon to organize a team. I said, hello Dr. Hawass, I'm from Hong Kong, you know, only his mind is probably, you know, look, another tourist. He nearly pushed me out, you know. But, I'm a very persistent person. Dr. Hawass had many offices. I knock on the door of this one. Knock on the door of that one. No one would go there 400 times. But I'm that kind of person. And finally, he could see my passion. Never give up! (See "The Robot, the Dentist and the Pyramid: Ancient Egypt Documentary (2020)," timecode 15:52, on Ancient Architects) (Sibson 2020)

Note: Following this clip, others that immediately follow will be from the same video but I will just list the timecode.

It was decided that a competition would be held between two teams, one from Singapore and one from Leeds in the UK, who would demonstrate the operation of their robot in a test shaft that was built out in the desert. A group from the Faculty of Engineering at Cairo University; Ali Radwan, a professor of Egyptology at Cairo University; Sabri Abdel Aziz, head of the Pharaonic Sector of the SCA; and Hisham El Leithy from the SCA were assembled to supervise and witness the competition.

One of the criteria for succeeding in the competition was that the robot would not leave any traces along its path traveling up or down the shaft. The previous robots had tracked mechanisms for movement, like those of a tank or a bulldozer, and had left scratches on the shaft walls and ceiling. The successful robot, Djedi, built at Leeds University under the direction of Professor Robert Richardson and Shaun Whitehead, was able to demonstrate a different mechanism that left no marks. Rather than gripping the ceiling and floor with track-like mechanisms, Djedi's motion was provided by employing mechanical actuators with rubber caps on the end on both sides of the front and back carriage. With the front actuators extended and gripping the walls and the rear actuators retracted, a gear known as a pinion that is meshed with a linear gear known as a rack rotated and advanced the rear carriage along a precision linear rail, similar to those used in machine tools, thus closing the distance between it and the front carriage. When the extent of its travel had been reached, the rear actuators were engaged and the front actuators retracted, after which the pinion gear reversed direction and the front carriage was pushed up the shaft.

A modern example of the use of a rack and pinion is the steering wheel in a car. When the steering wheel is turned, a pinion causes linear motion of a rack to push or pull the linkage that is attached to the front wheels, causing them to turn in one direction or the other.

In the same documentary in which Dr. Ng appeared, Shaun Whitehead gave an interesting perspective on the significance of the Queen's Chamber shafts, using models to illustrate his point. His first model was a gray foam wall block with the shaft cut completely through a corner of the block.

Holding the model, Whitehead says, "If I was to walk into an ancient room and find a hole in the wall, even if it disappeared up into the heights of the pyramid, that wouldn't be exciting."

Whitehead then picks up a small rectangular piece of the same material type as the model wall block he was holding, places it in the corner of the model at the beginning of the shaft and says, "If I walked into a room and found there was a block covering this hole that I could easily remove, then that would be a bit more exciting."

Whitehead then picks up a different model with the chamber wall facing the camera, so that there is no indication of a shaft. At the same time, he rotates the model, showing the details of the shaft cut through but stopped five inches (12.7 cm) short of the inside wall, and

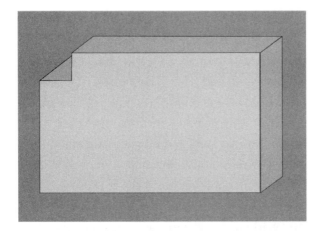

Figure 6.1.
Whitehead
model A.

Figure 6.2.
Whitehead
model B.

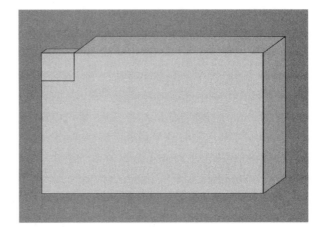

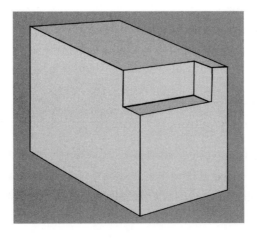

Figure 6.3.
Whitehead
model C.

says, "But if I came into the pyramid and I found that it appeared to be solid and only later discovered that it was hiding a shaft . . . that's very exciting. So somebody has gone to an awful lot of work to cut away this section of block, while leaving some material to hide the shaft. So, the question is, why hide this shaft when a human being can't fit up it?" (timecode: 7:56)

The answer to Shaun Whitehead's question is that the designers and builders of the Great Pyramid were not hiding these shafts. The termination at the Queen's Chamber end of both shafts was a crucial part of the design of the system. As were the so-called doors at the upper end of the shafts. Without these features being there, the Queen's Chamber would not function as intended.

The Djedi robot was equipped with a camera that had the ability to rotate and look in different directions, even to look behind itself and capture images of what is behind Gantenbrink's door. The images captured by the Djedi camera provide even more confirmation for my theory regarding the purpose of these shafts. The pins were first discovered by Rudolf Gantenbrink in 1993, but how they were fitted in the door was beyond Gantenbrink's ability to reveal, as well as out of reach of the Pyramid Rover's lights and camera in 2002. The image of the back of the door is incredibly revealing and unexpected.

The Djedi team explained what their robot revealed, to the best of their ability, in a documentary of the expedition that was produced later. Shaun Whitehead is the camera designer from the company Scoutek in Melton Mowbray, UK, and Robert Richardson is an engineering professor specializing in robotics at Leeds University. Their comments on the significance and meaning behind their discovery start at timecode 37:34.

WHITEHEAD: There were beautifully made loops on the back of the stone, which helps us try to work out what they were there for.

RICHARDSON: So, we were able to look around and see that it was a fully enclosed cube space. There was no shaft going off to the left or right, for example. And it was probably the floor that was most interesting.

WHITEHEAD: We were looking at red ochre marks that were not natural. The red markings are an enduring mystery. And opinions have ranged from over a wide variety of meanings. And that's part of the excitement. There's still more to discover. There's still more to learn. And the shafts and the marking are very much a part of that.

RICHARDSON: We have discovered a cuboid space, around 200mm × 200mm × 200mm. Experts in Egyptology, anthropology, symbolism, civil engineering, religion, members of the public have all looked at the findings. Nobody is sure what the red markings mean. Nobody knows why the small chamber was hidden behind the blocking stone, and nobody knows what the pins, loops and blocking stone are for, and nobody knows why the shaft entrance at the bottom was hidden.

On May 25, 2011, an article describing the Djedi robot's images was published by *New Scientist*: "As the camera can see around corners, the back of the stone door has been observed for the first time, scotching the more fanciful theories about the metal pins, says camera-designer Shaun Whitehead of the exploration company Scoutek, based in Melton Mowbray, UK. "Our new pictures from behind the pins show that they end in small, beautifully made loops, indicating that they were more likely ornamental rather than electrical connections." (Hooper 2011)

The article in the *New Scientist* magazine varies slightly with Whitehead's version presented in the documentary. It's a small difference, but holds significance to me because it indicates that the Djedi team knew about the theory that the pins were electrodes, and as I am the originator of that theory, it behooves me to respond to its dismissal by Whitehead by pointing out the following:

The metal loops can be seen to reenter the stone a slight distance below where they emerged. The features of these metal loops at the back of the "door" are consistent with the appearance of electrodes in a battery. It appears, also, that the pins have a ring of insulation around them, with the anode (the one on the right when looking from behind the door) seemingly less affected by chemical reaction. This ring of insulation is also evident on the front of the door.

Questioning why these electrodes are insulated from the limestone, one possible answer could be that if the limestone door became saturated with chemical, electrical flow through the saturated stone from one electrode to the other would be prevented. If the electrodes were not insulated, when the fluid level in the shaft dropped, the circuit may not have opened until the limestone had dried out.

A significant detail to highlight is the difference in the condition of the pins that are visible from inside the shaft from the condition of the metal on the back side of the blocking stone. I would argue that when these pins were first installed in the door, the condition of the metal visible from inside the shaft and behind the block looked the same. They were straight and round when first installed, but as time and use took its toll, the electrodes serving their purpose began to corrode due to the rise and fall of a corrosive liquid. This resulted in a taper to the pins that reflected the length of time the pin was immersed in fluid. The corrosion was less at the top of the pin, as this area was not submerged for the same amount of time as the bottom of the pin. See color plate 19.

Assuming we accept that chemical corrosion caused the pins in the South Shaft to wear in the way they did, a logical question would be, how long would it take for these pins to wear away and need to be replaced? When Upuaut II provided us with images for the first time of the condition of these pins, the one on the left had eroded to the point that a section of it had broken off and was lying on the floor of the shaft. This was not the result of any human intervention; the metal had failed and a section had broken off through being subjected to a corrosive chemical over time. The pin on the right was still intact and remained so until the Pyramid Rover expedition in 2002. Nothing was mentioned about it, but Pyramid Rover's first images of the door showed that it was the same as when Upuaut II photographed it. However, when the echo impactor, used for measuring the thickness of the door, was mounted on the robot and it returned to the door, the right pin was shortened and the echo impactor was able to fit beneath the shortened pin and press against the door. The phrase "Asking for forgiveness is sometimes more productive that asking for permission," comes to mind.

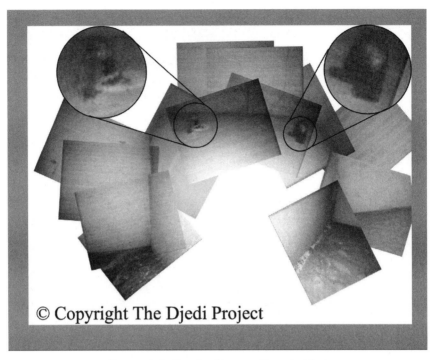

*Figure 6.4. Patched image from behind
Gantenbrink's door showing the looped pins on the back.
Image credit: Robert Richardson, PhD (the Djedi Project).*

MASONS OR MAINTENANCE ELECTRICIANS?

The continuation of the pins on the back side of the blocking stone was not the only discovery captured by the new robot. The Djedi robot also transmitted images of symbols and a line painted in red on the floor behind Gantenbrink's door. Peter Der Manuelian, an Egyptologist at Harvard University and director of the Giza Archives at the Museum of Fine Arts, Boston, said, "Red-painted numbers and graffiti are very common around Giza. They are often masons' or work-gangs' marks, denoting numbers, dates, or even the names of the gangs" (Hooper 2011).

Regarding the marks pictured in Figure 6.5, I propose that we look at them not as mason's marks, but as electrical symbols indicating the polarity of the electrodes (the red line) and the wiring of the electrodes.

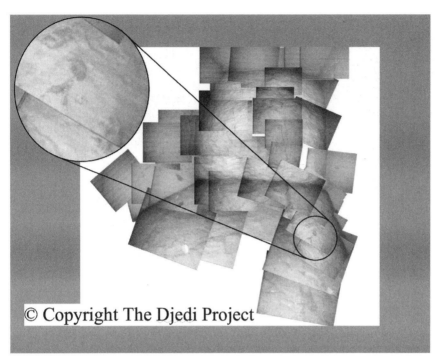

Figure 6.5. Patched image from behind Gantenbrink's door showing the symbols and a line painted on the shaft floor.
Image credit: Robert Richardson, PhD (the Djedi Project).

- The upper symbol appears to be shaped like the number 5.
- The center symbol that shows a round circle with a forked line below it could represent a cable with a twisted pair of wires that separate in order to be connected to a positive and negative electrode. It is positioned between the upper and lower symbols on the floor.
- The lower symbol is roughly similar to the upper with the exception of the top bend, which could go right or left, as when installed it wouldn't matter which way the loop was turned.
- All the symbols, including the line (which would logically represent the main cable coming from the USO) were more than likely positioned on the side that would identify the positive electrode. This is supported by the appearance of a greater amount of corrosion on the loop on the side where the symbols appear.

Figure 6.6. Cable with two wires to connect with contacts around metal pins.

When they were discovered, there was no immediate explanation for what these red symbols meant, but they are a significant discovery and have the potential to open an entirely new area of research in gaining an understanding of ancient Egyptian engineering symbolism. When considered along with the metal pins, the symbols provide a tantalizing possibility for how they might have been used to guide a maintenance worker. They explain the electrode use of the pins, and give us, I believe, motivation for further exploration in the future. The ancient Egyptians left us with the physical evidence that supports such an endeavor by providing us with an electrical schematic that showed how Gantenbrink's door and, most likely, the North Shaft "door" metal pins were wired. Even this evidence, existing in a small hollow within the mass of pyramid stone, far removed from other features, is connected to them and critical for the electron harvester's operation.

The images clearly show that the builders were aware that the pins would have to be replaced periodically and had made accommodations for performing maintenance on them. The quickest and most logical way to do that would be to create another blocking stone identical to the one currently installed, complete with new uncorroded pins and connectors, then remove the existing block and install the new one. After the exchange, the new one could be rewired according to the electrical symbols painted on the floor.

While it appears that the maintenance crew did not clean up after performing repairs in the space behind Gantenbrink's door, the debris was not the only evidence they left there. They also left instructions on how to wire the pins. The pins were probably tapered at the end, which allowed them to enter the S-shaped loop and push it open so that the loop gripped the pin. The vertical leg of this connector is not to scale (as very few wiring diagrams are) but the actual connector probably had a longer vertical leg up to the point when it is bent at right angles toward the center of the block.

To test my theory regarding the pins in Gantenbrink's door serving as a fluid switch in the Great Pyramid electron harvester, I created a model out of plexiglass and equipped it with electrodes that had a loop on the end. The process included creating a straight pin, forming the loop, inserting the straight pin through a hole in the door, and seating the loop end into a blind hole. Following that and using a torch to heat the copper, I bent the pins down 90 degrees against the door.

Judging by the amount of force I needed to bend the copper pins I had made for my model, it is now obvious that the thickness, or

Figure 6.7. Construction and successful test of Gantenbrink's door model.

diameter, of those in the pyramid's South Shaft would have to have been substantially larger when originally installed. Considering that the left electrode had failed during use and a section of it broke off without, presumably, human intervention, it would not have survived the force necessary to bend it when it was first installed in the "door." Without wear or corrosion, they would not have broken after use or, as observed, during the iRobot expedition before the blocking stone's thickness was measured.

When everything was assembled, the ancient Egyptians' level of design sophistication immediately became apparent. Once the pins on the shaft side were bent by 90 degrees with the looped end inserted into the hole, the pins were locked in position and could not rotate.

The most important discovery is the design of the two pins. Judging by their relationship to the size of the space, the pins are approximately 5/16 inch (.8 cm) in diameter. Figure 6.4 shows the metal looping around, with an apparent gap where the loop on the right pin seemingly disappears into the limestone block. The left pin shows signs of corrosion, similar to those in the main shaft, though not as severe. There also seems to be a white deposit around the left pin and its hole, while the right pin has what appears to be a black ring encircling the hole that penetrates through to the main shaft.

What was revealed by the Djedi robot describes an electrical device that was accessible to workers for maintenance. Considering the erosion on the pins in the main shaft (the negative electrode having broken off in antiquity) and considering the extreme tapering that was more than likely caused by the rise and fall of a corrosive liquid, another significant conclusion is that these electrodes must have been replaced periodically. At the same time, the electrical cables were probably replaced, and some of the shielding was left in the space. The pins were made so that they could be removed easily and others put in their place. However, a more efficient method would be to have a replacement block with new electrodes already fitted to drop into place after removing the existing one.

I should note here that, as with the entry location of the chemical supply shafts, my design for the wiring of the Gantenbrink's door electrodes is speculative, and, similar to the location discovered in the

Queen's Chamber North Shaft, when future expeditions are able to access and physically examine the area around the electrodes, they may find a completely different design. Whatever that design turns out to be, I have no doubt that it will make sense.

From the buildup of salt on the walls of the Queen's Chamber—up to one inch (2.54 cm) thick—to the design characteristics and other evidence discovered in the shafts, to the "noisome odor" assailing early visitors, everything about this chamber speaks of chemistry. Whether the chemistry is that of Drejewski, Cohen, or another preferred formula, there is no denying that only when analyzed through the prism of chemistry do we start to reveal answers to a millennia-old mystery.

ACCESS TO CONDUCT MAINTENANCE

This analysis and the questions it raises seem to indicate that there is a way to access the end of the Queen's Chamber shafts and that at least one additional passageway exists within the pyramid. This may lend support to Jean-Pierre Houdin's internal ramp theory, in which he proposes a ramp was constructed to enable the transport of pyramid blocks as the pyramid grew in height, and might suggest that the ramp was also used to access not only the Queen's Chamber shafts, but also other parts of the structure—whether they are working parts or not. The question that remains is where did the maintenance crew enter the pyramid to perform maintenance? Is it possible that there is a passageway behind the limestone blocks of the Great Pyramid that were shown to have a thermal anomaly on the northeast side, or did the maintenance workers take a shorter route by removing designated casing stones high up on the south side to reveal a horizontal passageway, with the end of the Queen's Chamber South Shaft a short walk away?

Besides prompting further exploration in the Great Pyramid, it is extremely exciting to find ancient symbols and be able to clearly connect them with a previously unknown aspect of human activity in millennia past. To discover that, as I had predicted in 1998, a possible chemical supply inlet shaft does exist, although not in the place I predicted it would be, provides more proof that the foundations upon which the

Queen's Chamber chemical reaction theory are built are even more solid than before. The multiple new discoveries opens a whole new area of study using knowledge and tools that have previously been excluded from ancient Egyptian studies. Fortunately, we are on the threshold of a new era, in which Egyptian engineers and scientists are recognizing that the Great Pyramid is more than a tomb. With the perspective of modern technical specialists and a renewed interest by Egypt's younger generation in solving the mystery of the pyramids, our thinking and attitudes about the capabilities of the ancient Egyptians will undoubtedly spread and their ancestors will eventually receive the credit they deserve.

SEVEN

SCIENCE OR PSEUDOSCIENCE? THAT IS THE QUESTION!

Thanks to the work of Ahmed Adly and other Egyptian researchers, such as Yousef Awyan and Mohamed Ibrahim, the Egyptians are waking up to the reality of what their monuments represent—that they are a historical record of their ancestors performing incredible feats of engineering by developing highly efficient, sophisticated tools. It has been quite common in Western literature to dismiss the abilities of indigenous peoples' ancestors and give credit for what they created to outsiders.

THE FOUNDATIONS AND STRUCTURE OF EGYPTOLOGY

Egyptians are also examining and questioning the foundations upon which the stories told about their history are based. The organized archaeological study of ancient Egypt, for the most part, was conducted by Europeans. In other words, the history books regarding ancient Egypt were written by outsiders, and the foundations for what followed were laid in the Victorian era. An awareness of how heavily Egyptology has been influence by Western scholars who absorbed the traditions and

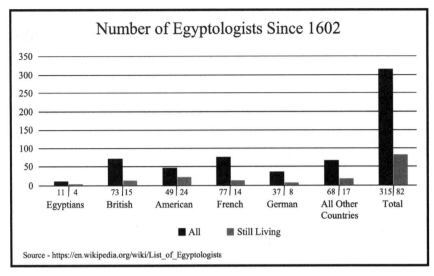

Figure 7.1. Egyptologists listed on Wikipedia.

attitudes that surrounded them as they performed their studies is evident when we look at the role Egyptians have played since Europeans first began to study Egypt.

German Jesuit scholar, Athanasius Kircher (1602–1680), is considered to be the founder of Egyptology. Wikipedia lists 315 Egyptologists from the time he performed his studies until today. During that period, 11 Egyptians are listed. Of those alive today, only 4 Egyptians are recognized, and they are overshadowed by 78 from countries outside Egypt. Figure 7.1 provides information on the countries these Egyptologists are from.

Out of the 82 living Egyptologists of different nationalities listed on Wikipedia, the four Egyptian Egyptologists are:

- Hussein Bassir, born 1973
- Zahi Hawass, born 1947
- Naguib Kanawati (Egyptian-Australian), born 1941
- Mahmoud Maher Taha, born 1942

However, there are many more Egyptian Egyptologists who apply their knowledge and training with dedication, perseverance, and care, not just in Egyptology but in other disciplines too. For instance, Aliaa Ismail, PhD, majored in both architecture and Egyptology at Cairo University and is gathering 3-D data with laser scanners within the tomb of Seti in the Valley of the Kings. (See ScoopEmpire 2017, "Meet Aliaa Ismail, Egypt's First Female Egyptologist.")

There are many more graduates from Cairo University who are blessed with a vast amount of knowledge that they teach to visitors who are attracted by Egypt's monuments and history and wish to learn more by visiting individually or as part of a tour. Those whom I have met personally, I have found to be incredibly kind and respectful. To say that all Egyptologists think the same way as those who receive most of the spotlight on the world stage would be a mistake.

The face of modern Egyptology to most of the Earth's inhabitants today is most likely Zahi Hawass. He is probably the most famous Egyptian Egyptologist of all time. This is mainly due to his uncanny ability to make noise and attract the media, which in turn attracts millions of people around the world. His passion and dedication to Egypt is unimpeachable, and under his leadership the wealth that has flowed into the country in the tourism industry alone is incalculable.

However, of the many beliefs and narratives that influence our understanding of our ancient roots on this planet, there is one narrative that I invite the reader to examine with a degree of skepticism. The stories that started to arise in the Western world regarding Egypt's past, told by scholars, astronomers, engineers, and philosophers from the Victorian Era to the present, need to be seriously questioned, I find. The main question focuses on those who control ancient sites—such as Egypt's pyramids, temples, and other sites—that are increasingly being recognized as clearly displaying the application and use of advanced technology.

QUESTIONING AUTHORITY?

Atlantis Rising (Doug Kenyon, editor) published an excellent article entitled "The Big Void" in its June–September 2018 issue. Its author is world-renowned geologist and author Robert Schoch, PhD. Professor Schoch serves as a full-time faculty member at the College of General Studies at Boston University and was a recipient of its Peyton Richter Award for interdisciplinary teaching. He earned his PhD in geology and geophysics at Yale University in 1983. He also holds an MS and MPhil in geology and geophysics from Yale, as well as degrees in anthropology (BA) and geology (BS) from George Washington University. Armed with significant scientific credentials himself, Schoch observed in his article that Egyptology, being related to archaeology, is more focused on history and art than on science, though in some cases scientists are employed to provide specialized information, such as determining the age of an artifact using carbon 14 or other dating methods or examination of terrain and structures using ground-penetrating radar or cosmic ray detection (muography). Schoch wrote:

> One of the issues seems to be that Egyptologists do not welcome the contributions of outsiders to their field. The majority of Egyptologists are (based on my experiences and observations) openly hostile toward any non-Egyptologists—especially scientists such as physicists, geophysicists, geologists, chemists, astronomers, and so on—who dare apply their expertise to Egyptological subject matter. Much of the problem may be that most Egyptologists are not scientists. Egyptology is traditionally more akin to such fields as art history, social history, linguistics, and the like. Many people have the mistaken idea that all Egyptologists are archaeologists, and archaeologists are scientists. The truth is that some (but not all) Egyptologists do carry out excavations and utilize archaeological techniques, and thus can genuinely be referred to as archaeologists, but it is debatable whether or not archaeologists should be referred to as scientists, or if it is often more a matter of archaeologists utilizing systematic techniques that on a superficial level appear "scientific" along with bringing in undoubted scientific expertise when it is useful to them (for instance, sending samples to laboratories manned by specialists—undoubted scientists—for radiocarbon dating). Here I will contend that many archaeologists, and certainly

most Egyptologists, are not scientists. This is not to say that their fields of study are not scholarly, or are not legitimate academic pursuits but, rather, to delineate distinctions among various disciplines. (I have great respect for many of my university colleagues who are historians, for instance, or philosophers, or ethicists, and so on, even though they are not scientists.) (Schoch 2018)

Interest in Egypt and the mysteries that still exist there even after, or because of, the uncovering of fabulous artifacts, has continued unabated for centuries. In 1993, Schoch was invited by independent Egyptologist John Anthony West to conduct an examination and provide an analysis of the weathering of the enclosure wall and structure of the Great Sphinx, which is located to the west of the village of Nazlet El Samman in front of the Giza Plateau where the Great Pyramids are located. I have visited this village many times and have lifelong friends there. From a professional geological perspective, Schoch's analysis concluded that the Sphinx enclosure and the body of the Sphinx were subject to water erosion due to rain runoff from the Giza Plateau that must have occurred over a long period of time. Schoch is a trained scientist and was using methods and training that would need to hold up to scrutiny from his fellow geologists. Overall, with few exceptions, his peers understood and accepted his conclusions. The reader can access his books and published articles on his website, The Official Website of Robert M. Schoch.

Schoch's analysis revealed that the currently accepted dates for the construction of the Great Sphinx could not be correct because of a lack of rainfall during the time period suggested by Egyptologists for its construction. He was met with a firestorm of opposition from Egyptologists, the most vocal of whom was Zahi Hawass, who at the time was director of the Giza Plateau. A casual observer would assume that two professionals with PhD after their names would surely have equal status and therefore be on equal footing in a debate. Schoch, however, calls into doubt the balance of authority in this particular instance by stating that, unlike himself, Egyptologists cannot be considered scientists. The obvious question arising from that conclusion is who do we believe: the scientist (Schoch) or the Egyptologist (the latter being considered by Schoch to be a nonscientist)?

Exploring this further, we know that the title of doctor is applied to some professions that, while performing immensely valuable work and benefitting society, cannot be considered to be hard sciences. Take, for instance, the doctor of divinity degree that is bestowed on theologians. A vastly knowledgeable person holding this degree may be faithful to the scientific method in their vocational or avocational pursuits, but by its nature their doctorate cannot be considered to be a branch of the hard sciences.

DESCARTES'S PHILOSOPHY

Other than taking Schoch's word for this opinion, does other evidence exist to support his claim? Certainly, the main public figures of Egyptology hold passionate beliefs, to the level of religious fervor, that they are providing the public with truths about Egypt's past. But could it be that they are acting more from an ideological perspective than from the results of hard, scientifically acceptable methods employed in their work? The beauty of René Descartes's philosophy and his famous meditation on doubt, "That which can be doubted must be rejected," is that passionate beliefs do not hold up if they cannot be supported by the objective application of the scientific method—which guides hard sciences in their pursuit of truth and should really guide everyone.

> But inasmuch as reason already persuades me that I ought no less carefully to withhold my assent from matters which are not entirely certain and indubitable than from those which appear to me manifestly to be false, if I am able to find in each one some reason to doubt, this will suffice to justify my rejecting the whole. And for that end it will not be requisite that I should examine each in particular, which would be an endless undertaking; for owing to the fact that the destruction of the foundations of necessity brings with it the downfall of the rest of the edifice, I shall only in the first place attack those principles upon which all my former opinions rested (Descartes 2008).

Following Descartes's philosophy, therefore, if a single piece in a structure, or mode of thinking, can be doubted or cannot be proven, the whole structure must come down and be reconstructed with pieces that are unimpeachable. You could argue that this rarely happens, and you may be correct, but in the hard sciences it has become accepted wisdom and essential. Not so by at least one Egyptologist, however, who angrily and unprofessionally attacked Schoch when he offered a geologist's analysis on the weathering within the Sphinx enclosure.

> The initial hoopla peaked in February 1992 at a "debate" on the age of the Great Sphinx held at the Chicago meeting of the American Association for the Advancement of Science. As the New York Times put it, *"The exchange was to last an hour, but it spilled over to a news conference and then a hallway confrontation in which voices were raised and words skated on the icy edge of scientific politeness."* Egyptologist Mark Lehner could not accept the notion of an older Sphinx, personally attacking me by labeling my research "pseudoscience" (Schoch 1992 and 2010).

While the old proverb "People in glass houses should not throw stones" is frequently ignored, it is useful to remember when criticizing another person for whatever reason. Considering how Robert Schoch has been mistreated by Egyptologists in the past, particularly by American Egyptologist Mark Lehner, who accused him of practicing pseudoscience, or Zahi Hawass calling new ideas that challenged the established beliefs of Egyptologists the products of "hallucinations," it is understandable that Schoch would write a negative review of Egyptologists' scientific standards, but are his criticisms merely sour grapes for the way he has been treated or is there justification for them, with evidence to support him?

It would seem that even a layperson with a rudimentary understanding of how science works and its associated rules of evidence would agree that a more powerful argument supporting Schoch's contention that Egyptologists are not qualified to be considered scientists is made by Egyptologists themselves. What better source of testimony exists than the words from three world-renowned Egyptologists' own mouths in September 2002 as they were peppered with awkward questions from the inquisitive host and narrator of a documentary chronicling the exploration of the Queen's Chamber South Shaft and much-heralded penetration of Gantenbrink's door?

Struggling to provide satisfactory answers to the narrator's questions, the objectivity and scientific thought process of leading Egyptologists from Egypt (Zahi Hawass), the United States (Mark Lehner), and Germany (Rainer Stadelmann, 1933–2019) were on full display and broadcast to the world on September 16, 2002, from the Giza Plateau and, particularly, inside and outside the Great Pyramid. The video of this broadcast can be seen on YouTube in "Into the Great Pyramid" (National Geographic 2002). I will insert the timecode in my text below in brackets.

The 2002 live broadcast reached an audience of 30 million in the US and many millions more around the world. Of these, there were probably a few million that recorded the program on their VHS recorders. One of those millions was me. While I recorded it for future viewing, I probably only watched the last 10 minutes over again before the video went into storage and eventually became obsolete. Considering the evolution of modern technology over the last 20 years, when this video became relevant to my research again, I didn't go digging around in a plastic storage container, even though I still have my old VHS player (also in storage and not hooked up). I turned on my computer and searched for it on YouTube. What a difference in viewing experiences! The benefits of searching for a relevant scene and also being able to read the live transcripts is a dream compared to what was available in 2002.

As the National Geographic hosts, Laura Greene and Jay Schadler, built up the anticipation and excitement with the help of Zahi Hawass and Mark Lehner, Greene held a revealing interview with Hawass in the Queen's Chamber:

GREENE: One question I have to ask you, Zahi—people back home are gonna wonder—has anybody had a peek? [12:48]

HAWASS: No. But I have some ideas, we will talk about it tonight. But I don't think that anyone could find out exactly what is behind this door. [12:52]

Greene then passes off to Schadler at the overseer's tomb in a segment that lasts for around eight minutes before the filming returns to Greene and Hawass in the Queen's Chamber discussing what was appearing on the computer screen as the Pyramid Rover ascended the shaft. As if a beat wasn't missed since her previous question of Hawass, Laura Greene persisted:

GREENE: No, Zahi, as we have said, you are the expert. The question on everybody's mind has got to be, what do you think is behind the stone door? [26:20]

HAWASS: Laura, listen. We are sure now that this door is about three and a half inches thick. Therefore, it's a door with two copper handles for the rope to pull that door. Something should be behind it. [26:29]

GREENE: What though? What do you think it might be? [26:42]

HAWASS: I will tell you, if we will be optimistic and think the treasures of Khufu could be hidden here, and even if we are not optimistic and see a small corridor, or a hollow or something behind it, I believe it's a great discovery, because something inside the Great Pyramid to be seen for the first time; it's important for everyone to see that. [26:45]

Following preliminary discussions on the iRobot's function, its demonstrated ability to drill a hole through limestone, and some of its struggles ascending the shaft, Hawass and remote sensing expert Meg Watters climbed up the south face of the pyramid and paused at a spot about halfway up where the Queen's Chamber South Shaft should have emerged had it continued at the same angle to the outside. Their conversation went as follows: [1:20:25]

HAWASS: We are near that shaft now.

WATTERS: OK.

HAWASS: But the archaeologists here at Giza looked everywhere on the surface, OK, on the south here and could not find any evidence at all of this shaft anywhere that has a hole outside.

WATTERS: Can you tell me what the shafts were for? [1:20:43]

HAWASS: It is my theory, my belief, that the south shaft functioned as a symbolic corridor for the soul of the King, as the Sun God Ra, to go through that shaft—that in the mind of the Egyptians, the soul of the King will go through that shaft. The northern one is also a model corridor for the King's soul, as the God Horus, to go in the north hemisphere. [1:21:08]

NARRATOR: The role of the shafts must have been very important, because building them was incredibly difficult. They weren't just drilled through the pyramid. As we have seen, they were carefully laid out as the pyramid went up layer by layer. So why would the King's soul need these special shafts to exit the pyramid instead of just simply going down the Grand Gallery and using the regular exit? Egyptologist Rainer Stadelmann thinks that the King refused to. [1:21:41]

During the voiceover narration at the time the question is raised, respected German Egyptologist Rainer Stadelmann, the director of the German Archaeological Institute in Cairo from 1989 to 1998, is seen climbing up the Ascending Passage and standing beside the North Shaft inside the King's Chamber. He responded to the question:

STADELMANN: This was very important because, as written in the Pyramid Texts, the King said that 'It is my distaste to descend into the Earth. I, the King, ascend directly to the sky.' [1:21:59]

Understandably slightly perplexed by Stadelmann's answer, another awkward question was posed.

NARRATOR: But if the King's spirit went from his burial chamber to the sky, then why go to the trouble of building shafts for the

Queen's Chamber, where the King was never buried? The answer may be linked to magic. Mark Lehner thinks the Queen's Chamber was a dummy chamber, inhabited by the King's soul that would need to move out of the pyramid to the heavens. Then wouldn't a stone door pose a problem? [1:22:27]

These were all excellent questions that would be on the minds of any objective viewer, but then the narrator introduced the idea that magic was involved, which led to American Egyptologist Mark Lehner providing an answer.

LEHNER: A blocked passage would be no problem for the King's soul. It didn't have to be open for the King to ascend. The principle means for going into the otherworld in any normal Egyptian tomb was what we call a false door. A dummy door. But when they made a simulated door out of stone, then it's magical and then it can commune with another world.

This raises the question, if the King's soul can pass through stone without any problem, why does he need a shaft? Does he need directions on which way to go, and the directions are written on sacred scrolls behind Gantenbrink's door? It's all very confusing. Four shafts from two locations pointing in different directions that are necessary additions for the King's soul to exit his tomb. Four shafts, the construction of which would require enormous amounts of time and resources to include in the design scheme. Absent any objective evidence to support the notion, this portrays the scientific mindset of at least three world-renowned Egyptologists. But there is more:

NARRATOR: But if the blocking stone is kind of a magical false door, why does it appear to have handles? [1:22:55]

Again, another excellent question, the answer to which is provided by Egyptologist Rainer Stadelmann, who is still stationed in the King's Chamber by the North Shaft:

STADELMANN: These copper fittings are magical instruments which the King can take, the soul of the King can take, when it ascends.

It uses them to open this block and pass through the stone to the sky. [1:23:11]

There is a slight difference between Lehner's and Stadelmann's explanations. Lehner claimed that the soul of the King would have no problem passing through this blocking stone, while Stadelmann stated that magical handles are necessary for the soul of the King to raise the block before passing through the stone to the sky, even though the shaft does not continue on to the sky. The narrator continues her ruminations in voiceover:

NARRATOR: Magic was essential to Egyptian rituals, and Khufu's burial chamber must have contained all sort of papyrus scrolls with magical writings. Prayers to help him on his way to the heavens. But what about behind the stone?

STADELMANN: It might be possible that there is, let's say, a papyrus roll behind these stones. There's no place for much more.

Stadelmann's comment is very revealing. "There's no place for much more." How does he know that? At this stage in the exploration, according to Hawass, nobody has taken a peek, even though he said that when a peek is finally taken, there may only be a small space behind it. Hawass and Stadelmann have essentially told the audience not to expect much of a discovery when the robot reveals to them for the first time what is behind the so-called door, while Hawass, for one, is claiming that nobody has taken a peek. It should be noted, moreover, that there is ample room behind Gantenbrink's door for a huge space, using the dimensions between the end of the shaft and the outside and also if the builders wanted to take up a large amount of room to the east and west. Surely, a room the size of the so-called King's Chamber would fit—which is what they were hoping to find and which was the impetus for this exploration.

When the camera was inserted into the hole that had been drilled through the "door," great excitement and bated breath permeated the Queen's Chamber. Finally, the camera revealed a small space, as suggested by Stadelmann and Hawass, and to much fanfare, Hawass announced that across the space was another "door" with a crack! A very important discovery and important for the whole world to learn about as it was revealed for the first time!

HAWASS: We are here in front of a discovery. This is incredible and I am really happy that we got this. We found the space. We found another sealed chamber. [1:24:13]

NARRATOR: This is incredible. This is the first time a space has been advanced within the Great Pyramid within 130 years.

HAWASS: I know, but Laura, this is very important. This is something I'm very proud that finally we revealed the first mystery of the Great Pyramid of Khufu. [1:25:41]

During this documentary, Egyptologists essentially confirmed Schoch's assertions that Egyptologists are not real scientists. But did they also shade the truth about not knowing what was behind Gantenbrink's door before it was broadcast live to millions of viewers around the world? Was what we witnessed on September 16, 2002, an honest account of the application of science in pursuit of truth, or were Egyptologists following a script given to them by filmmakers who just wanted to tell a good story—a story presented in the same tradition as other stories told through the lenses of filmmakers, whose main objective throughout history has been to attract the largest audience possible for their creations? In this case they succeeded, for by all accounts tens if not hundreds of millions of people tuned in.

Perhaps the words of a respected Egyptologist who was not involved in this production stated it best. Iorwerth Eiddon Stephen Edwards, CBE, FBA (21 July 1909–24 September 1996) was the Assistant Keeper in the Department of Egyptian and Assyrian Antiquities from 1934–1955, and Keeper of Egyptian Antiquities from 1955–1974. In his illustrious forty-year career, he was considered to be the leading expert on pyramids, and his book *The Pyramids of Egypt*, published by Penguin in 1947, is considered to be a classic scholastic work that has been through many revisions over the years. In an interview for a documentary for Netherlands Television and reissued by History's Mysteries under the title "Secrets of the Pyramids and the Sphinx" (History's Mysteries 1995), Edwards gave his thoughts about the importance of pyramids.

> Edwards: Pyramids, in general, are not very popular with Egyptologists because I think they are too plentiful. There are about a hundred of them. And also, I think they have acquired something of a bad name because they have attracted so many cranks. (As of 11/28/2015, "Secrets of the Pyramids and the Sphinx," Maximum Truthseeker, YouTube, timestamp 4:30.)

For those who are not familiar with the term "crank," as used by Edwards in this case, a list of synonyms are: character, codger, crack, crackbrain, crackpot, eccentric, flake, fruitcake, head case, kook, nut, nutcase, nutter [British slang], oddball, oddity, original, quiz, screwball, weirdo, zany. Hmmm . . . I think I see "fruitcake" in there.

It is understandable that a researcher, using the best scientific methods, may interpret the symbols and writings of an ancient culture and come to an understanding that the ancient culture was deeply vested in a belief in magic. This culture's belief may have ruled their lives and formed a basis for most decisions that they made. The question is, do real scientists include magic when explaining their observations?

Not being immersed in these teachings and beliefs, an outsider may question an Egyptologist's reliance on magic to explain physical features crafted into the Great Pyramid. In fact, one might argue that the Egyptologists involved in the 2002 exploration of the Queen's Chamber South Shaft in the Great Pyramid, while they may be well-meaning and sincere in their beliefs, have revealed by their performance a breathtaking vacuum where objectivity, science, and common sense should exist. Invoking magic to explain physical features found within the shaft reveals a mind filled more with fantasy and religion than a scientific mind. Fantasy and religion may have their place in the minds of believers. However, it should be recognized for what it is and be absent from scientific analysis.

Nevertheless, so powerful is the grip of magic on Egyptology that any attempt to introduce a new idea or different way of looking at something is met with resistance and, in a lot of cases, hostility, such as when Lehner accused Schoch of practicing pseudoscience.

Inspired by the observations of R. A. Schwaller de Lubicz that the erosion of the Sphinx and its enclosure was the result of water flowing down from the Giza Plateau over 10,000 years ago, John Anthony West had encouraged Robert Schoch to get involved, for, if the science supporting the theory for an earlier date for the Sphinx construction was to be debated, it was required that the debaters hold a PhD. West, who did not hold a PhD, was not qualified to advance his de Lubicz-inspired theory in the debate, which was organized by the American Association for the Advancement of Science. In the same video in which Edwards made his comments, journalist Paul William Roberts, who covered the 1992 debate, made the following compelling observation:

West is really an academic's worst nightmare. Because here comes somebody from far out in left field with a thoroughly well-thought-out, well-presented, coherently described, beautifully written [theory] full of data you can't refute, and it pulls the rug from beneath your feet. So how do they deal with it? They ignore it. They hope it will go away. But it won't go away. [32:13]

Ancient writings as well as recent writings are not the best evidence. Without firsthand testimony as to the origins of and meaning behind the writings, what exists may be considered hearsay. Most of what we have learned about ancient Egypt and the Great Pyramid is based on writings from the past. With respect to what its origins are, what its purpose was, and how it was constructed, what is painted or inscribed on an ancient wall may have the same historical value as graffiti that is spray-painted on the side of a train or the bridge it passes over.

In *The Demon-Haunted World: Science as a Candle in the Dark*, Carl Sagan said, *"The cure for a fallacious argument is a better argument, not the suppression of ideas."* Who could reasonably argue with this point of view? Ideas that are supported by fallacious arguments have always been a factor for some when shaping a message to convince people to buy what is being sold. Politicians are well known for it. If the first thing a person did before writing anything online was to learn what logical fallacies mean and how they are used, they will be doing themselves and the world a favor.

Totschweigetaktik, an Austro-German word, means "death by silence." Invariably, when a different idea that is free of logical fallacies is proposed, for many reasons it will be resisted by silence. In "The Big Void," Robert Schoch identified a condition that is common not just in the culture of Egyptology but also in other fields of endeavor around the world. It is known as NIH or the idea that something was "Not Invented Here" and is therefore ignored.

> It is not always clear to me why the Egyptian authorities, and certain non-Egyptians working with them, continually want to downplay, politicize, and even dismiss or totally deny the discoveries of well-trained scientists and experts. There seems to be a pattern that if something is not "discovered" by the card-carrying Egyptian Egyptologists or their close associates, then it must not be real—or if the reality of the discovery cannot be denied, then it must not be significant and/or it was known all along. (Schoch 2018)

The only objective scientists in the Queen's Chamber during the course of this explorations, it seems, were the Pyramid Rover builders and operators. They were using a product designed and built out of components that were underpinned by hard science. There was no ambiguity or any reason to question their confidence that their robot would work, and if it didn't, they would employ science and engineering to fix it.

Hidden secrets, mysteries and answers to questions that have persisted for millennia have an appeal that is hard to ignore. But even after the plethora of books Egyptologists have published on the subject of pyramids, there remains to be explored mysteries and secrets that attract millions of visitors a year. Some internet conspiracy theorists have argued that Egyptologists have discovered some remnants from a lost civilization that they are not sharing with the world. I would not argue support for that point of view, but I do recognize that keeping valuable information secret is not new. It flourished in full force in Italy when tile layers would hold the tricks of their trade within a closed group or guild. Italian tile layers were the finest in the world and they wanted to keep it that way. During that period in our history, who could blame them? The question is, will this practice survive the scrutiny of future generations?

Edward Leedskalnin was a Latvian immigrant who built Coral Castle in Florida and performed amazing feats that have drawn millions from around the world to marvel at them. Unfortunately, Leedskalnin was afflicted with the Italian tile layer syndrome, and the secrets that would allow others to replicate his feats are still a mystery that he took to his grave (Dunn 1998, 109).

Nevertheless, even without help from holders of secret knowledge, physics, science, and engineering continue to evolve and share their knowledge with the world to improve living conditions and enhance awareness through their inventions. I wonder if we would be further along if they had revealed the secrets they hold? I doubt it. With the exception of Edward Leedskalnin, I suspect the answer is a resounding no.

In a remarkably well-crafted monologue at the close of the documentary about the opening of the "door," Jay Schadler provides a philosophical, though necessary, caveat for what we had just been treated to:

Wah, what a night, ah? Already the work of recording and analyzing what was found here has begun at the site, and we should tell you that from, really from the moment this project was conceived, everyone here at National Geographic has understood that we were involved in a kind of twenty-first-century experiment. A blending of hard science and television. They can make strange bedfellows, and no one knows that better than us. After all, science demands patience—television craves experience. Still, both are seeking a kind of revelation to reveal what is hidden beneath the surface of things and to be honest, I suspect that no one's heart beats faster at the moment of discovery than the scientist's. If we have managed tonight to give all of you at home a glimpse into that world, into that moment when the unknown becomes known, then we'll consider it a success. Tonight, we still stand on sacred ground. Home to the world's first great civilization. Now, for Laura and Dr. Hawass and everyone here at National Geographic, I'm Jay Schadler. Goodnight from the Giza Plateau and the Great Pyramids.

The mystery and the magic spells that have been fondly enriched and amplified before being served to a ravenous population that covers the globe has its roots in old Hollywood movies. Even the drama surrounding Lord Carnarvon's death after a shaving cut became infected included speculation that it was a mummy's curse that sent him to his grave. Were ancient evil forces really at work when *Death on the Nile* was being filmed?

Egypt, with its pyramids and temples, has long been the greatest show on Earth, and hushed utterances about magical mysteries and "secrets to be revealed" reside in the scripts of many promoters who wish to attract seekers of wisdom and enlightenment. The magic and the mystery is an alluring potion that captivates and enchants. But if we focus on just the science, even more wonders can be revealed.

> Science arouses a soaring sense of wonder. But so does pseudoscience. Sparse and poor popularizations of science abandon ecological niches that pseudoscience promptly fills. If it were widely understood that claims to knowledge require adequate evidence before they can be accepted, there would be no room for pseudoscience. But a kind of Gresham's Law prevails in popular culture by which bad science drives out good. (Sagan 1996, 6)

Gresham's law, referred to by Sagan, originated in the finance industry, where it is said that bad (overvalued) money drives out good (correctly valued) money by appealing to less discriminating investors.

With what we have witnessed from world-renowned Egyptologists, does Sagan's metaphoric use of Gresham's law, where bad science drives out good science, apply to their work product? Considering the lack of scientific evidence put forward to support conclusions explaining important features in the Great Pyramid, there is a strong argument in support of Schoch's opinion, which argues that Egyptologists are not scientists, despite the number of skeptics that may align against him. I have found that most ardent self-proclaimed skeptics seem only to be interested in focusing their skepticism on new ideas. That's a shame. They end up defending the status quo while attacking true skeptics, like Schoch, who have legitimate questions and innovative ideas about what has previously been accepted as truth.

Referring again to Arthur C. Clarke's quote that *"Any sufficiently advanced technology is indistinguishable from magic"* and considering at least three eminent Egyptologists' inclinations to invoke magic when trying to explain features of the Great Pyramid, would it be outrageous to suggest that perhaps they had become entangled in Clarke's maxim? Were they interpreting what they were studying as magic while in truth they were faced with "sufficiently advanced technology" but were unable to recognize it for what it is?

However, many today do recognize that the Great Pyramid was constructed using advanced technology for the purpose of creating energy in a unique and technologically advanced way. The truth, however, can only be found through the application of good, scientific, Cartesian principles. It is a candle in the dark where magic will find no comfort.

EIGHT

SAQQARA'S HIDDEN TREASURES OF ADVANCED TECHNOLOGY

A popular tourist attraction when traveling to Egypt is Djoser's pyramid at Saqqara—commonly known as the Step Pyramid. When I visited there in 1986, I was completely unaware of another site called the Serapeum just over a half mile away. There is not much to see from a distance, but when you get closer you will see steps leading down an incline to heavy iron doors set in the bedrock. Behind those doors are probably the most mysterious and confounding discoveries ever made on this Earth. An account of my research prior to 2010 is published in *Lost Technologies of Ancient Egypt*. In this chapter, I will broaden the scope of my previous studies, which leads to the question, "Why was the Serapeum created?" A possible answer to this question will shock you.

WHERE ARE THE MACHINES?

Since I ended *Lost Technologies of Ancient Egypt* with the question, "Where are the machines?" it seems appropriate to ask the same question in this book. Its relevance persists in many minds that consider the conclusion both Petrie and I reached—that the ancient Egyptians

Figure 8.1. Christopher Dunn inspecting the remarkable precision of a granite box in the bedrock tunnels of the Serapeum.

Figure 8.2. Steps leading to the Serapeum at Saqqara.

Figure 8.3. Entrance to the Serapeum at Saqqara.

used sophisticated machines to shape stone. The Serapeum's monolithic boxes reinforce this point of view.

When passing through the iron door, you enter a cavernous room. Figure 8.4 shows a large granite block that is roughly finished but ultimately destined to be crafted to serve as a lid. Its weight is estimated to be around 28 tons. To the right of the entrance is a tunnel, which, partway down its length, contains a box that has been roughly quarried and roughly finished on the inside (see Figure 8.5). Passing by the lid in the entrance hall, to the right is another long tunnel in which large openings with vaulted ceilings are cut into the walls and staggered along the length of the tunnel. These openings are cut above and below the level of the tunnel floor and are commonly referred to as crypts. Inside the crypts have been placed massive granite boxes that measure approximately 13 feet long × 7.5 feet wide × 8 feet high (3.96 × 2.28 × 2.44 m). Each crafted from a single block of granite, with the insides cut to shape, but without their lids, the boxes have been estimated to weigh 70 tons and the lids 25 tons. (A complete survey of the entire facility needs to be performed in order for more accurate estimates of each individual box.) Significantly, the boxes are centered in the crypts and installed in a recess in the floor. Other tunnels exist in this complex and all told there are 24 such boxes in these underground bedrock tunnels.

Figure 8.4. (above) Ben Van Kerkwyk and Yousef Awyan in the Serapeum entrance hall by the granite lid.
Figure 8.5. (below) Rough-finish box in Serapeum tunnel.
Image credits: Ben Van Kerkwyk.

There is a drop of several feet from the tunnel passage floor to the crypt floor. Why this was necessary has long been a mystery. Installing the boxes would certainly have been less complicated and difficult if the crypt floor was on the same level as the tunnel floor, so why the creators of these enigmatic, subterranean vaults made it harder for themselves by going to all the extra work poses a question that I will try to answer. Their effort is to be admired, but what were they trying to achieve?

Engineers are well-known for stretching their ingenuity when faced with an engineering mystery. I'm fortunate in the past to have been in the company of some exceptional engineers while they have been grappling with this question. One idea, which I will address later in the chapter, rang true to me, as it opened a path to providing possible answers to most of the major questions raised, such as:

- Why was extreme precision needed on the inside of the box?
- Why was this precision *not* a requirement near the bottom of the box?
- Why did the lids to the box also need to be precise?
- Why were the crypt floors lower than the tunnel floor?
- Why were the boxes placed in a recess in the floor of the crypt?

As Ahmed Adly discovered, the inside of the box he measured was precise, with the walls of the interior flat and parallel with each other and the corners all being 90 degrees square. His measurements provided support for the observations I made in 1995, 2001, and 2013, with the exception that he used a laser measure and digital protractor, whereas I used a toolmaker's precision straightedge and solid square. The straightedge, precise to within two ten thousand parts of an inch (0.0002 inch [0.00508 mm]), provided an additional attribute: the flatness of its surfaces. At the location where I used the 12-inch straightedge, no light was seen passing through the interface where the steel straightedge was placed on the granite surface—regardless of how the straightedge was turned, whether vertically, horizontally, or at an angle.

From a manufacturer's perspective, these discoveries essentially *prove* the existence of a technologically advanced civilization in Egypt's remote past.

With young engineers now performing their own inspections of the boxes in the Serapeum, "Where are the machines?" is usually the first question asked by those who do not believe the ancient Egyptians were that technologically advanced because of what has been found in the archaeological record. What are the implications to accepting that they *were* that advanced and what we have been taught in school is *wrong*? The history of the human race would need to be reassessed and history books that place Egypt behind Greeks and Romans in terms of technical accomplishment would have to be rewritten. Is it surprising that there is resistance to such an idea?

The question "Where are the machines?" also presents a bit of a conundrum. On the one hand, we cannot assume that machines used in ancient Egypt would be similar to and work in the same way as modern machines. Supporting that point of view is an advanced machine from ancient Egypt that has the shape of a pyramid, and when you consider how it is designed and built and how it may have operated, there are no modern machines to compare it with. On the other hand, the pyramid electron harvester concept, as presented in my first book, *The Giza Power Plant,* is questioned because it does not resemble any machines that we are familiar with, and there are no known electrical devices or appliances in the archaeological record that it may have powered. The mindset that the ancient Egyptians were not capable of creating precise complex machines seems to be one of the reasons for not accepting that machines were used to craft many of the impressive artifacts found in Egypt. But yes, they were, and they did!

In the context of the Serapeum boxes, I am fully aware that if I am asked "Where are the machines?" pointing to the pyramid machine is not quite the answer the questioner was seeking. Nevertheless, the pyramid machine does provide evidence that the ancient Egyptians were familiar with machine-like precision and sophisticated function, even if the pyramid machine is not designed to finish-grind the inside of a granite box.

While my peers who are familiar with manufacturing processes and what tools and methods are capable of achieving might look at the magnificent boxes in the Serapeum and conclude that precision machines *had* to have been used to make them, a layperson without a similar background probably feels justified in asking, "But where are the machines you speak of that did this work?"

In *Lost Technologies of Ancient Egypt*, the foreword by mechanical engineer Arlan Andrews Sr. presented a cogent argument that the question of lost technologies is not confined to the technologies developed by prehistoric civilizations, but that technologies used even in our lifetime are vulnerable to obsolescence when those who developed them and knew how to use them eventually pass away. What they created does not necessarily continue to exist in its originally manufactured form but is disassembled and the materials repurposed for use in modern products. It could be that the car you drive today, or even the bridge you drive it across, contains materials that were once used to manufacture an earlier model automobile.

I recently discussed with Andrews the difficulty some people have accepting that machines *must* have been used to create many of the artifacts in Egypt, and he kindly provided me with a quotation to be included in this book:

> To those critics who demand that the proposed ancient machine tools be discovered before paying any attention to the incredible manufacturing technologies displayed by the dozens of "crypts" at Saqqara, let us imagine finding a similar structure in the future, only in the deserts of Mars, not Egypt. What would the reaction be? What would be the most important questions?
>
> If those early interplanetary explorers were to come across a cliff opening that was the entrance to a long, carved tunnel which opened to a series of enclosures housing incredibly huge black stone boxes weighing many tons, precisely machined on the insides and situated in recessed niches, yet having no workers' tools left behind, do you think that their first reaction would be "Where are the machines?"
>
> Obviously not! The scientists and engineers would report the impressive artifacts assuming that the application of machines was a given: "They are there. They are massive. They are obviously manufactured. They are precisely machined, because no conceiv-

able human (or Martian) hand (or tentacle) could possibly accomplish the flatness, the parallelism, or the plethora of other features we have documented, unless they possessed sophisticated guidance mechanisms, for cutting tools along with associated metrology instruments.

Our Martian research team's relevant questions then are, "What advanced civilization made these monumental artifacts? How did they do it? When was it done? What was the purpose? And, most importantly, where do we go from here with our research?"

To my mind, present-day researchers of Egyptian artifacts like those at Saqqara should focus on the available evidence—the boxes. They are massive. They are obviously precisely machined because no conceivable human hand could possibly achieve their geometry and exactness without the assistance of precisely powered machinery.

So, the relevant questions today are the same questions our hypothetical future Mars explorers would have: "What advanced civilization made these monumental artifacts? How did they do it? When was it done? What was the purpose?"

And the most relevant question for today's researchers is: "Where do we go from here?"

(Email received from Andrews on January 21, 2022)

Andrews described an alternative path that discoverers may follow, depending on their training and background and the background and training of those who join them in analyzing the discovery. In the fictional account above, perhaps because of a planet-killing asteroid impact or coronal mass ejection (CME) event, the entire civilization on Mars was completely wiped out. The planet became uninhabitable and there were no survivors.

A well-used aphorism claims, "Those who forget history are doomed to repeat it." I would like to challenge that caution by asking, "Whose version of history are we supposed to believe and not forget?" Historical interpretations of ancient artifacts, such as the boxes in the Serapeum, do not include a detailed engineering analysis using modern instruments. The activities that are recorded and reported as historical events come from many sources, including prime witnesses who were on the scene when a discovery was made; scholars who examine ancient artifacts; books that provide interpretations and stories about their meaning; other observers, philosophers, and reporters; and, ultimately, the general public.

The first visitors to the hypothetical Mars tunnels most likely will have a background in the sciences, being perhaps a geophysicist, geologist, or even an engineer. Along with others with a technical background, it's the kind of group you would expect to be selected to crew a space mission to another planet. I think we can agree that Egyptologists did not have a hand in creating the Serapeum, and they would probably not be selected to crew a space mission to Mars.

The first explorers who discovered the Serapeum made their discoveries before the Industrial Revolution began or picked up full steam. The discovery was made before Egyptology was established as an academic study, but after it was, their analysis of the site was not informed with the knowledge and education that a spacefaring civilization would need. That is not a criticism but a reality. Nonetheless, recent interdisciplinary research teams possessing a broad range of knowledge have begun researching the site and finding that a different story needs to be told.

The difference between Earth and Mars in the two scenarios is the extent of devastation on the planet where they are located. In Andrews's fictional scenario, we can imagine that the planet Mars suffered a mass extinction event and there were no survivors. On Earth, one of the effects of a mass extinction gave rise to what is known as the "bottleneck theory." This is when the diversity in human DNA is reduced significantly when a large part of earth's population is wiped out. One event, that reduced the world's population to just a few thousand, is believed to be when the super volcano, Toba, erupted around 75,000 years ago: "In 1998, the bottleneck theory was further developed by anthropologist Stanley H. Ambrose of the University of Illinois Urbana-Champaign. However, some physical evidence disputes the links with the millennium-long cold event and genetic bottleneck, and some consider the theory disproven" (Wikipedia, Toba Catastrophe Theory).

While the Toba eruption is one cause of mass extinction, others such as asteroid, meteorites, and coronal mass ejections have been identified as causes of others. It doesn't take much of a stretch of our imagination to consider what might have existed before such events occurred. How technologically developed were the inhabitants? How much of their infrastructure would survive such a conflagration? What has survived for thousands of years that have caused a lot of us to scratch our heads are the pyramids and the many other examples of monolithic stonework being worked to a high order of precision. I believe that one such extinction event brought an advanced civilization to a screeching halt, leaving few survivors who had the resources to understand and continue that society's advances in technology and who did not possess the ability to rebuild.

Perhaps, like us today, those who created the boxes in the Serapeum were served by a scientific and technical infrastructure that was a delicate web of knowledge and materials skillfully orchestrated to provide them with comfort and convenience. Today, a significant event could wipe out one or more critical components necessary for our lifestyle to continue. Our comforts and convenience would be like a house of cards that could collapse entirely if we suffered such an event in the future.

In our era, the study of ancient artifacts falls under the purview of archaeology. In Egypt, it is controlled by the Ministry of Antiquities. It is often said that the winners get to write history. In the case of the Serapeum, the winners were and are those who planted their flag first. The Egyptologists on Earth got to write their opinions about the monolithic boxes in the tunnels at the Serapeum first. In the alternate fictional scenario on Mars, the scientists and engineers got there first and the story they would tell was completely different and more than likely the one that would endure. However, engineers are now entering the tunnels of the Serapeum with a different set of tools than those used by Egyptologists. Instead of taking with them a brush and trowel, they are employing precision instruments such as straightedges, squares, protractors, and laser measuring instruments to gain a deeper understanding of the truth inside the boxes' empty spaces.

A few years after *The Giza Power Plant* was published, it became clear to me that it was not making—and would not make—a dent in the way mainstream Egyptologists think of ancient Egyptian history. I had fallen victim to *totschweigetaktik* (death by silence), and came to realize that if my observations were going to be accepted and make a difference in the status quo, the people who had the most to gain would be the Egyptians. If Egyptian history was going to be rewritten, then the Egyptians should decide if it should be rewritten, and the Egyptians should be the ones who rewrite it.

I also concluded that if I wanted to penetrate the cracks in the structure erected by Egyptologists, *The Giza Power Plant* was, metaphorically, the thick end of the wedge. In 2010, I published *Lost Technologies of Ancient Egypt*, which was designed to be the thin end of the wedge. The people most suitable for hammering that wedge into the cracks would be Egyptian engineers. To that end, I dedicated the book to "The Egyptians and their glorious heritage." Also, within the book I appealed to Egyptian engineers to examine for themselves artifacts I identified as products of a highly advanced civilization that possessed superior manufacturing abilities—their ancestors. In *Lost Technologies of Ancient Egypt,* I wrote:

> "The complexity of a box is not quite that of a statue or even the cornice block near the Valley Temple at Giza. Yet the message from the tunnels of the Serapeum is one of metrology and exactness. While at the moment precision is inferred from the geometry and symmetry of the statues, the boxes in the Serapeum let us know quite clearly that the ancient Egyptians were no strangers to such exactness. *The lapidary work on the insides of these boxes should invite further studies by Egyptian engineers* who can interpret the information that is crafted into the meticulousness with which these giant, enigmatic artifacts were made."

That strategy seemed to be the right approach. Egyptian engineers are now interested in examining with modern tools the monuments built by their ancestors, tools Egyptologists may not be aware of. As it turns out, accompanying them now are Egyptian tour guides who interact with tour groups interested in *Lost Technologies of Ancient Egypt*. These groups are populated by engineers from all over the world who visit Egypt to see for themselves, while at the same time pondering if they can add value to Petrie's and my work by making their own discoveries and adding their own voices.

While in Egypt in September 2021, I began to learn the extent to which *Lost Technologies* has impacted the younger generation. Many new media personalities, particularly YouTubers, are now drawing visitors to Egypt to examine its artifacts and the evidence of advanced civilizations. On September 28, 2021, I had a meeting with Hany Helal, former minister of scientific research and higher education and lead coordinator of the ScanPyramids project, and Hamada Anwar, an international law judge who is representing Egypt's legal interests in the project. Both gentlemen were extremely kind and intelligent, and I came away relieved that there is a new breed of government official involved with the pyramids, officials who have different attitudes toward new ideas. Both have received copies of my books and expressed an interest in cooperating further.

Following my flight home, and after returning to my office to continue writing this book, I contacted a Facebook friend, Ahmed Adly, an Egyptian engineer who has done his own studies of Egyptian artifacts. I was interested in learning what his views were regarding the opinions of his peers in Egypt and if he would provide me with a quote for my book. He said he would be happy to do it and sent me a report. After reading his report, which is included in the foreword. I watched his YouTube videos and was very impressed by his presentation style, his use of graphics, and his accuracy of reporting. Considering that he is probably the most credible Egyptian researcher on social media, I am grateful that he agreed to write the foreword for this book.

Ahmed Adly's YouTube channel has several excellent videos on the subject of *Lost Technologies in Ancient Egypt*. Together with the UnchartedX YouTube channel, Adly's videos are probably the best out there, in my opinion, as they are based on Adly's genuine engineering knowledge and skills. His presentations are very thorough and objective and are becoming increasingly popular, with some attracting a million viewers. Currently, they are in Arabic, but he has told me that his plan in the future, as time allows, is to create English translations. What I learned in Egypt, combined with Adly's report, was significant. My overall impression is that there is a quiet grassroots revolution of young Egyptian professionals who are rejecting the stories of Egyptologists and are now more interested in learning about the actual lost technologies of their ancestors. They have rejected the idea that the pyramids were tombs and are willing to consider alternative, more scientifically aligned explanations for their existence. There is a strong willingness to accept the idea that the pyramids were machines.

Some comments regarding *Lost Technologies of Ancient Egypt* have expressed disappointment that I did not provide my own concept of what such machines would look like and how they would operate. To an average person, when they think of a machine, they think of large pieces of metal. Forgings and castings come to mind. They are machined and assembled with precision to perform actions that affect the condition of other materials. The heavy, cast-iron frame of a milling machine can be clearly identified as part of a machine, even if much of what was originally attached to it has disappeared. Rusting in scrap yards across the world are metal objects that were once parts of machines that now sit patiently waiting for natural forces and time to slowly consume them or to be repurposed and shaped into other products. After how many years will they no longer exist? Five thousand? Ten thousand?

THE MACHINES OF THE SERAPEUM

It's impossible to convey the mindset that has been shaped by over a half century of work experience in manufacturing, especially if a significant portion of that time has been spent in performing the manual and mechanical activities involved in creating precision products. One cannot achieve it by reading books or watching YouTube videos, no matter how entertaining they may be. So I'm not sure if words are sufficient to provide you with the full opening of awareness I experienced when discovering the precision that was crafted into not just the boxes in the Serapeum but also Khafre's pyramid, where, upon placing a precision straightedge against the inside surface, I exclaimed, "SPACE-AGE PRECISION!" A comment for which I have been heavily criticized and sometimes mocked. Those who criticize and mock are not seeing what I see. They see an empty space, a wall of granite. Their vision ends at the surface of the granite. For a machinist/toolmaker/engineer like me who has devoted tens of thousands of hours to the creation of precise objects, any precise object that I come across can be viewed as a window to a world where knowledge, materials, and machines are orchestrated into manufacturing processes that produce a flow of products for demanding customers. The average person who uses these products is unaware of the complex yet specific range of activities that were necessary to create the products they interact with every day. The only performance they are interested in is the performance of their purchases—the product in their hand (smartphone) or the product that surrounds them (car, train, or airplane). In reality, the entire orchestration of manufacturing activity is important to them, for it affects the quality and price of their purchase.

When I first discovered the precision with which these monolithic boxes were crafted and noted that there was an unfinished box near the entrance to the bedrock tunnels, I proposed that the monoliths were finish machined and polished underground before being installed in the crypts. In this way, changes in temperature were avoided that might affect the granite's precision if it was finished aboveground and then brought below. This level of precision is not crafted into an artifact unintentionally. It is a design characteristic that is important to the function of the artifact, and efforts are made to ensure these characteristics of precision are not subject to change. Reviewing the work of other researchers, such as Ben Van Kerkwyk of the UnchartedX YouTube channel, who was guided by Yousef Awyan, a talented Egyptian sculptor, a different picture is emerging of the full scope of activity that was taking place thousands of years ago. This picture is similar to what takes place in modern manufacturing.

Walk into any manufacturing plant and you will find a product that is in various stages of production. A rough forging or casting may be seen in one area, and in another area the rough forging or casting will have been rough machined. All of its surfaces may not be required to have the same texture and precision as the final product but will be left with enough material (stock) for the final machining to remove in order to achieve its required dimensions, tolerances, and finish. It is in what is known as a semifinished state.

The final product may have evidence of forging, rough machining, and ultraprecise machining, with the forged, rough, or not so precise regions left alone, as they are less important for the performance of the product. For instance, every surface on a forged crankshaft is not semifinished or finish machined. The crank webs and counterweights generally are not machined and keep their original forged finish.

The three states of manufacturing to remember when trying to find reasonable answers to what exists in the tunnels of the Serapeum are:

1. Rough. (As created on site or delivered by the foundry, mill, forge, or quarry.)
2. Semifinished. (As created on site or delivered by the foundry, mill, forge, or quarry.)
3. Finished. (Precisely finished on site.)

It is now revealed that boxes in the Serepeum display all three states of manufacturing. There is a rough box and lid, there are boxes contained within crypts that are semifinished and are without lids, and there are boxes with lids, such as the one I inspected in 2001, that are finished to a high order of precision. A professional survey of the entire installation by qualified engineers is required to create a more complete picture of what the Serapeum contains and what it means.

Following the path a piece of raw material may take through various processes involving workers and their machines, the last step in the process will see the finished product inside an inspection laboratory populated by highly skilled and knowledgeable quality inspectors along with a variety of metrology instruments and machines that will be used to verify every important attribute crafted into the product. It is within this environmentally controlled area that we find part of the answer to the central question of "Where are the machines?"

In a corner of the inspection laboratory at Danville Metal Stamping, a company I was employed by for 27 years, is a large machine known as a CMM (coordinate measuring machine). The manufactured product is placed on an inspection-grade granite surface plate, and the machine is programmed to navigate three-dimensional space to determine the exact coordinates of specific features on the product using a touch probe, or laser, in order to compare the physical object with the model designed by the computer. But it is not just the manufactured product that enjoys the feel of smooth, flat granite against its surface. The bridge that carries the Y- and Z-axis travels along the X-axis, which also needs a surface upon which to travel that is flat and precise.

It is important to know that a section of the CMM's granite base also serves as a way along which a machine carriage travels. This machine carriage is equipped with air bearings and, essentially, "floats" on a thin film of air. In other words, a single piece of granite has a dual purpose in this machine. It serves as an ultraprecise reference plane upon which products are placed while at the same time serving

Figure 8.6. One of Danville Metal Stamping's CMMs.
Image credit: Gardner Peck (Danville Metal Stamping).

Figure 8.7. Example of machine where ways are
separate from the work table.
Image credit: Gardner Peck (Danville Metal Stamping).

as a precise way for the testing machine. This is an important function that traditionally has been separate from the work table upon which products are positioned and secured.

Before getting into the details of how this relates to the Serapeum, I would like to discuss critical information I have learned over the years about the nature of granite. This is information that a countertop installer will probably not share with you because it is not relevant to how your granite countertop looks or functions. However, it is highly relevant when crafting and maintaining calibrated precision surface plates used in manufacturing.

The granite surface plates used in manufacturing plants, such as Danville Metal Stamping, require frequent calibration by independent professionals who are contracted to inspect, resurface, and certify that the surface plate conforms to industry standard ASME B89.3.7-2013.

John Barta of Barta's Precision Granite Co. in Cleveland, Ohio, is one of these professionals, and he provides these services to companies across the Midwest. When I was working in Indianapolis, I had the opportunity to talk to Barta and pick his brain about his experience with granite. One of the most important aspects of the nature of granite that he shared with me was how it is affected by temperature. What he had learned in his career has importance to the discussion in this chapter.

Granite Is Sensitive to Heat

John Barta told me of a case where he visited a shop that had a surface plate that had somehow developed a slight bulge or crown in the center. A slight bulge could mean only microns but was enough to swell the granite so that its flat surface did not conform to required specifications. The cause of this condition, it turned out, was a single incandescent light bulb positioned above the plate. While I don't recall him saying how close the lightbulb was to the surface of the granite, what its wattage was, or how long it had been turned on, this is a subtle but important example of why large granite surface plates require an even temperature throughout the material.

When a granite surface plate manufacturer receives a piece of granite from the quarry, the granite is semifinished to machine-capable tolerances, perhaps to within .001–.005 inch (.0254 –.127 mm) and is placed outside until there is a customer's order. In this uncontrolled environment, granite is subject to changes due to the heat of summer and the cold of winter. It could be in the yard for years before an order is received, at which time it is brought into a temperature-controlled environment for final finish.

As I discussed Ahmed Adly's research in the Serapeum in this chapter, I emailed it to him to make sure I had represented him honestly and correctly. I was very pleased with his response and grateful to learn that a PhD candidate is conducting studies in Egypt on the effect that heat has on Egypt's granite monuments. He writes:

> When you mentioned the swell of the granite, it brought me to another. Around 6 months ago, Mrs. Rabab Abd Elhakim, a teaching assistant and a PhD student in the Faculty of Arts, Geography Department of Mansoura University in Delta, Egypt sent me her

Master's Thesis, with Title "Thermal extremes and their effect on Ancient Egyptian monuments." In one of the chapters, she carried out an interesting simulation experiment to study the effect of heat on color and size of the stones by bringing several rock types, mudbricks, limestone, diorite, granite . . . and other types that were commonly used in ancient Egypt, putting them in a closed Venticell oven for 12 hours continuously, setting it to 50° C (simulating Egypt weather). With measures before and after the heating process, she recorded an expansion in the stones and rocks (Mudbricks: 100 microns, Marble: 10 microns, Diorite: 10 microns); for other stone types, no changes were recorded. However, in a later phone call with her, she mentioned that longer periods are expected to have an impact on the form and size of granite, and she addressed many evidences in her thesis from Komombo temple and many other sites in Upper Egypt specifically due to its hot weather, where several granite statues and blocks slowly expanded then its surfaces started to crack and disintegrate. Yes, this point was overlooked by many, thanks for bring it up. (Ahmed Adly, email to author, June 15, 2022)

Granite Has Memory

The granite that was being held in the surface plate manufacturer's yard has changed shape many times through seasonal hot and cold cycles and needs to adjust to the temperature required for the use of laboratory- or inspection-grade surface plates before final lapping can take place. This is where, for want of a better word, its "memory" comes into play. After being brought inside to a controlled environment and after sufficient time, the granite will return to the same shape it was before being placed outside.

And this is where it gets interesting. In more ways than one, the precisely crafted modern CMM sitting in a manufacturing plant in the American Midwest shares its "DNA" with the millennia-old, precisely crafted monoliths of ancient granite housed in the mysterious, dusty bedrock tunnels of the Serapeum at Saqqara, Egypt. The mystery surrounding the question, "Where are the machines?" can be partly answered by assuming that the most important structures and foundations of the machines that applied the final finish to the inside surfaces of the monoliths in the Serapeum are *the boxes themselves.*

When a high level of precision is required in the final product, temperature is critical when transferring a finished article from one place to another. As with a piano that requires tuning after being moved, a surface plate needs to be recalibrated. With this in mind, the answers to the question regarding the kind of machinery that was employed in the finish machining and polishing of the monoliths become clearer, and all the evidence indicates that the granite boxes in the Serapeum received their final finish and calibration *after* they were placed in the crypts.

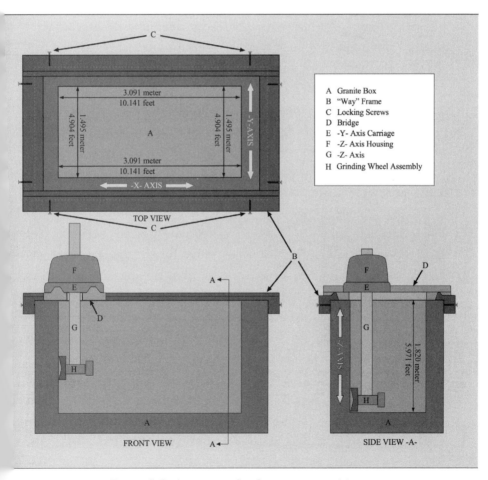

Figure 8.8. Anatomy of a Serapeum machine.

We know the Egyptians were capable of creating ultraflat surfaces that had no other features that would obstruct their lapping plates—the underside of the lids and the top surfaces of the boxes are evidence of that—but what about the inside surfaces of the boxes? The simple answer is that the top surface of the boxes could have been used as precision ways, like the CMM, upon which a superstructure comprising a bridge riding on bearings navigated along the length of the monolith.

Assume that this bridge was traveling along the X-axis. The bridge was equipped with another precision "way" upon which a carriage traveled in the Y-axis. A vertical structure mounted on this carriage, known as the Z-axis, was equipped with the tools necessary for grinding the interior of the box. The tool used to cut the surface may have been a circular grinding wheel that was able to be indexed in 90-degree increments around a rotary axis, typically identified as the A-axis. In this manner, the face of the grinding disk can grind each interior wall of the box. Also, the task of creating a square corner with a small radius where adjacent surfaces meet is easier to accomplish. However, a feature left in the bottom corner of one of the boxes seems to open the possibility that grinding tools with different geometries may have been used.

The corner at the bottom of this box, seen in Figure 8.9, is left with a feature that one would normally identify as a gusset that provides

Figure 8.9. Unusual feature in bottom of Serapeum box.
Image credit: Ben Van Kerkwyk.

extra strength between two or three tangent surfaces. However, it also poses several other questions regarding why it is there and how it was created:

- Does it reflect the shape of the tool that was used to cut the box?
- Observing that the adjacent surface does not appear to be perfectly flat, is the box in a semifinished state and will this feature be removed in final machining?
- Do other boxes have this feature in them?

Preliminary examination of the interior of these monoliths indicates that ultraprecision at the bottom was not required. In the box I briefly examined in 2001, the precise flat surface I noted near the lid is not present near the bottom.

There may be a reason for this if we try to answer the question, "Why were these boxes made so precisely?"

It has become clearer now that the ancient Egyptians were very advanced in their knowledge and use of sound and vibration and used that knowledge to affect their environment and way of life. A possible reason that it may have been necessary to install the monolithic boxes inside a recess in the middle of the crypts is that they were subject to vibration and, if not contained, they could move out of position.

If the boxes were stimulated to vibrate, then for what reason? Also, considering the geometry and precision of the inside surfaces, I'm compelled to ask, "Did the inside of the box serve as a resonant cavity?" An example of a resonant cavity is a laser, where parallel mirrors serve to amplify photons that are released when energized electrons are stimulated to return to their natural orbit around the nucleus of an atom.

Considering that the walls of the Serapeum box cavities are parallel and square to each other, what conditions would affect a medium within the cavity, whether that medium is a gas or a liquid? There is no denying the testimony of hundreds of tourists who have lain down inside the resonant granite cavity known as a sarcophagus inside the King's Chamber of the Great Pyramid. When the box is struck, or a particular frequency of sound is directed into the cavity, the person lying in it feels something they have never felt before. Recall the descriptions from my Facebook friends regarding their experiences (page 80).

There is no clear and definitive answer to the question, though the idea does shed some light on why the bottom of the cavity is not as precise as the upper section. Because the box is made from a single piece of granite, it has a thick bottom, and any vibration in that area would be dampened. Therefore, all the vibratory action would be in the upper part of the walls. Another reason why the bottom of the boxes' interiors may not have required the same care and precision as the upper parts is even more mind-blowing when considering the input from a variety of sources.

On my Lost Technologies of Ancient Egypt tour in 2018, I entered the Serapeum with my group, which included aerospace engineer Eric Wilson. His idea regarding a possible function of the boxes was that they were used to grow crystals. During the preparation of this book, I reached out to Eric to ask him if he still supported that idea. Remarkably, he responded that he not only still thought it was a possibility, but further insights as to what may have been going on there were shared with him by a New Zealand friend, Lauren Kurth, when they both visited the Serapeum in 2020. Lauren suggested that the boxes may have been surrounded by water that was heated to a certain temperature, and this would have helped with the growing of crystals.

I must admit, I was blown away by such an imaginative idea because it resolved other questions that I have had in my mind since first visiting the Serapeum in 1995. Eric and Lauren's insights provide a possible solution to the mystery of why the granite boxes are placed in a recess within a chamber that is cut below the surface of the tunnel floor. They also open a new path of inquiry into the manufacturing steps that were taken in crafting the boxes and the critical considerations of how precision can be maintained in granite if it is kept at a specific temperature.

TAKING THE PATH TO PERFECTION

- Specifications are given to the quarry to deliver a piece of granite with sufficient material left to remove in order to achieve final dimensions.

- Granite is received, brought into the tunnels, and receives semifinish machining. The granite may be precise to within 0.100 inch (2.54 mm) with material left for final finishing.
- The boxes are installed in the crypts prior to final machining.
- The lids are semifinished, with the exception of the underside, which is finished flat to a high level of precision.
- When surrounded by water of the correct temperature for an appropriate amount of time necessary for the granite to acclimate, the top of the box is finished flat, with all four sides being on the same precise plane.
- An attachment similar to the one illustrated in Figure 8.8 is securely mounted on the top surface, and the insides of the box are precision machined and lapped to achieve the specified finish and tolerances.

The box is now ready for use. The question is, "What was it used for?"

While working for an engineering company in Indianapolis in the mid-1970s, one of my projects was to manufacture an electrophoresis plating machine. It was used to apply ceramic material that was suspended in water to turbine blades, creating a thermal barrier. One of the features of the machine was a shaft with blades on the end that almost touched the bottom of the tank. These blades rotated very slowly, their purpose being to stir the material that sank to the bottom of the tank, keeping it suspended in the liquid and preventing it from settling on the bottom.

With the Serapeum boxes, we might be able to create all the necessary conditions to grow crystals. A large vessel is caused to vibrate but is prevented from moving due to being inserted in a recess in the floor. The large vessel is kept at an even temperature by surrounding it with water of the optimum temperature—which explains the crypt floor being below the tunnel floor. The questions that remain, though, are: "What kind of crystals, and what results were expected that required the vessel to have perfectly flat and parallel inside surfaces?"

ENGINES OF CREATION AND THE SERAPEUM

Past until Present-Day Manufacturing

Past and present manufacturing has been and is based, principally, on reductive processes. Material is made by combining elements in a furnace, the results are poured into molds, and various processes turn the material into different raw material products. The raw material is received by a manufacturer and then shaped by diverse means, some of which include the removal of material using specialized machines.

Present and Future Manufacturing

Currently, the introduction of computers and sophisticated software has, in some cases, allowed greater efficiency and results. What are known as additive technologies are now being used for creating products that were previously manufactured in the traditional manner. Some jet engine parts are now being made with a 3-D printer using a laser that fuses material that is continually added in discrete amounts in a precise location in three-dimensional space.

To travel further into the future, conceivably we are now beginning to explore the territory occupied by the creators of the Unidentified Arial Phenomena that dazzle and baffle us with their gravity-neutralizing, amazing demonstrations of flight. What advances in manufacturing did the civilization that conceived of and produced these machines make? It might help if we reach back into the past a quarter of a century and revisit the visionary work of K. Eric Drexler.

In 1986, Drexler authored a remarkable book titled *Engines of Creation: The Coming Era of Nanotechnology*, about which the following was written by the publisher: "*This brilliant work heralds the new age of nanotechnology, which will give us thorough and inexpensive control of the structure of matter. Drexler examines the enormous implications of these developments for medicine, the economy, and the environment, and makes astounding yet well-founded projections for the future.*" (Drexler 1986)

"OK," you may ask, "what does all this have to do with the Serapeum?" I would answer perhaps nothing—perhaps everything. At the risk of it being nothing, and the possibility of it being everything, I cannot ignore what Drexler said about how, using nanotechnology, an aircraft engine could be assembled one atom at a time by what he described as microscopic "universal assemblers."

Drexler's vision of how nanotechnology could be developed and applied covers a broad range of activities in the medical and manufacturing field. If you asked Drexler the question, "Where are the machines?" he would describe a machine so small you would not be able to see it with the naked eye.

Similarly, if Dustin Carr invited you to a concert that featured a forty-piece orchestra, on arriving at the concert hall you would ask, "Where are the instruments?" Carr's nanoguitar, created in his lab, is so small you would need an electron microscope to see it, and when played, the subtle energy and precision of a laser beam impacting the strings would cause them to vibrate at a frequency you would be unable to hear. If the rest of the orchestra were built that way, there would be no noise complaints from his neighbors.

Universal Assemblers

Drexler describes universal assemblers as "second-generation nanomachines" that are built on a microscopic scale to perform an additive manufacturing function that can build complex structures one atom at a time. Unlike a 3-D printing machine in which the grains of different elements are fused together, nanomachine universal assemblers would be programmed with all the information they needed to select a specific atom from the environment, bond with the atom, and transport and deliver it to a precise location where it is caused to bond with atoms already connected to the structure that is being formed. Drexler points out:

> Because assemblers will let us place atoms in almost any reasonable arrangement, they will let us build almost anything that the laws of nature allow to exist. In particular, they will let us build almost anything we can design—including more assemblers. The consequences of this will be profound, because our crude tools have let us explore

only a small part of the range of possibilities that natural law permits. Assemblers will open a world of new technologies. (Drexler 1987, 14)

In everything I have been describing, I have stuck closely to the demonstrated facts of chemistry and molecular biology. Still, people regularly raise questions rooted in physics and biology. These deserve more direct answers.

–Will the uncertainty principle of quantum physics make molecular machines unworkable?

This principle states (among other things) that particles can't be pinned down in an exact location for any length of time. It limits what molecular machines can do, just as it limits what anything else can do. Nonetheless, calculations show that the uncertainty principle places few important limits on how well atoms can be held in place, at least for the purposes outlined here. The uncertainty principle makes electron positions quite fuzzy, and in fact this fuzziness determines the very size and structure of atoms. An atom as a whole, however, has a comparatively definite position set by its comparatively massive nucleus. If atoms didn't stay put fairly well, molecules would not exist. One needn't study quantum mechanics to trust these conclusions, because molecular machines in living cells demonstrate that molecular machines work. (Drexler 1987, 15)

To have any hope of understanding our future, we must understand the consequences of assemblers, disassemblers, and nanocomputers. They promise to bring changes as profound as the industrial revolution, antibiotics, and nuclear weapons all rolled up in one massive breakthrough. To understand a future of such profound change, it makes sense to seek principles of change that have survived the greatest upheavals of the past. They will prove a useful guide. (Drexler 1987, 20)

To have any hope of understanding the full scope of the Serapeum and what it was used for, we must not only look into the past but project our minds into the future. Some people in the world are more adept at doing that than others. One of them is my friend, Arlan Andrews Sr.

Through our mutual interest in pyramids, I first met Andrews in Indianapolis when he was working for Bell Labs in the late '70s. This was during the time I was working on *The Giza Power Plant*. In 1986, Andrews excitedly introduced me to Eric Drexler's *Engines of Creation*, and we spent many hours discussing its ramifications.

In 1989, Andrews left Indianapolis for Albuquerque, New Mexico, to work for Sandia Laboratories, and while there, in April 1991, was selected by the American Society of Mechanical Engineers (ASME) to serve as a Fellow in the Technology Administration Office at the Department of Commerce in Washington, DC. Following this assignment, in January 1992, he joined the Office of Science and Technology Policy (OSTP), a group of scientists and engineers in the White House Science Office, which is housed in the Eisenhower Executive Office Building next to the White House. Their mission was to advise the president on emerging technologies and what their implications were for changes in society and, particularly, consequences that could impact national security. While there, in November 1992, he founded a science fiction think tank named SIGMA.

Figure 8.10.
Arlan K. Andrews Sr.
Image credit:
Arlan Andrews, ScD.

That same month, Andrews invited Drexler to attend a meeting with OSTP staff and the director of the National Science Foundation. Following that, he wrote the first White House endorsement of molecular nanotechnology for the report, Science and Technology—A Report of the President to the Congress 1993. In the chapter "Manufacturing— The New Competition," Andrews discussed future developments in aggregation (additive) processes—then called 3-D faxes and now known as 3-D printing; microelectromechanical systems—today's tiny robots; and molecular manufacturing—Drexler's nanotech concepts.

Inspired by Drexler's book on nanotechnology, Andrews wrote a short story for the science fiction magazine *Amazing Stories* (May 1989) entitled "A Visit to the Nanodentist," in which he described a world in which dentists were using nanotechnology to restore teeth to their natural state. The molecular machines were specifically designed and programmed to remove decay and rebuild the tooth with the correct material, one atom at a time, while at the same time the cells of the surrounding gum tissue were repaired. Carrying on the theme of future nanotech applications, Andrews's later novel *Silicon Blood* (Hydra Publications, March 2017) imagines a society where the same advanced technology is used to create a new drug cartel.

On March 17, 2017, an article appeared on CNBC's website titled: "Mini-nukes and mosquito-like robot weapons being primed for future warfare" (Daniels 2017). The next day, Andrews shared the article on his Facebook page, with the following comments:

In 1991, at a table in the cafeteria in the Old Executive Office Building, I was telling other Fellows and a woman guest about Eric Drexler's Nanotechnology Conference in California I had just attended, as a Fellow in the Technology Administration of the Dept. Of Commerce.

"What could that technology be used for?" my friend's guest asked. I replied, "Well, imagine a flying robot the size of a housefly zooming into the Oval Office a hundred yards from here, sitting on the wall with full video and audio sensors." She thought for a moment. "Send me a copy of your trip report, please."

After she left, I asked who she was, not recognizing her name. "Oh, she's the ex-wife of [a famous science fiction writer] and is with The Agency." A month later I was invited to a CIA symposium on nanoelectronics. (Arlan Andrews, Facebook post, March 17, 2017)

Thinking Outside the Box

In general, people who think outside the box are admired and encouraged for their ability to imagine creatively how to improve society. Once the fruits of their imagination are brought into reality, a new and different box is created that defines and controls the activity that sustains the replication of their creations.

One might think that when an out-of-the-box thinker defines the parameters of a new box, their new box is much larger than the one they stepped out of. As a matter of physical size, the out-of-the-box, redefined new box does not necessarily get bigger. Bigger is not always better.

Boxes within which aircraft and rocket engines are created today cover thousands of acres. Foundries, manufacturing plants engaging in specialized parts production, assembly plants, and warehouses dot the landscape across the planet. Within these boxes, sometimes known as gigafactories, are smaller boxes, virtual or physical, where humans and machines perform specific functions within defined parameters. The humans entering the plant know where to go and where their "walls" are located, and they perform their work within those walls, using the tools and training provided to them. Drexler writes: *"A sprawling system of factories staffed by robots would be workable but cumbersome. Using clever design and a minimum of different parts and materials, engineers could fit a replicating system into a single box—but the box might still be huge . . ."* (Drexler 1987, 55).

Drexler's vision for creating rocket engines using nanotechnology would shrink all those existing boxes that are necessary to produce the engine down to the size of a single large box. The 1970s prediction of how humanoid-robot-operated lathes would function in the future, and the reality of what exists today, are proof that predictions do not always manifest as they are described. In the case of the humanoid-robot-operated lathe, what came to pass was far greater than what was predicted. Will we say the same about Drexler's prediction that nanotechnology will transform not just manufacturing but every other field that influences human existence on this planet? It stretches the boundaries of imagination, and I cannot imagine it being outshone when it is brought into reality. But if history teaches us anything, it is that future generations will improve whatever we can imagine and create, with their own imagination and research, something that is far greater.

History has shown us that Drexler is on solid ground when he predicts that the machines we currently use to achieve an end result will be replaced by much smaller devices achieving greater results. In the early nineteenth century, Charles Babbage (1792–1871), an English mechanical engineer, conceptualized and created the world's first mechanical computer, and some consider him to be the father of computers. Babbage's difference engine and analytic engine were large machines made up of thousands of precision machined mechanical parts, such as shafts, rods, gears, cams, and pegs (Babbage et al. 2012). What these massive machines accomplished is now achieved using a mere fraction of the capabilities of the smartphones we now carry around in our pockets.

Let's Take a Walk on the Wild Side

The smaller, but still huge, box in which Drexler imagines that manufacturing can take place in the future contains raw materials and workers. The raw materials are the elements found in nature and their atomic content. The workers are assembler nanomachines that are programmed to select an appropriate atom and bond it with others within a predetermined structure. Before they have been set to work building a jet engine, their nano cousins have assisted materials scientists and engineers in creating new materials that are specific to providing the optimum operation and output of the engine they are building. These new materials provide both hardness and strength, with greater resistance to heat and corrosion.

The elements from which the jet engine is created are customized for the demands put upon each particular part of the engine. For instance, the hot section of the engine's single crystal combustor may be constructed using material that includes ceramic elements, such as aluminum oxide, silicon carbide, and tungsten carbide. The combustor is bonded to the rest of the engine without welds, bolts, or rivets, but transitions seamlessly with a crystal lattice structure made up of creative and functional pathways for efficient cooling and fuel delivery. Discussing the creation of an advanced rocket engine built by universal assemblers, Drexler notes: *"Compared to a modern metal engine, the advanced engine has over 90 percent less mass"* (Drexler 1986, 62).

When the engine reaches the end of its life cycle, it is quickly replaced with a new one and sent back to the box from where it came, where the nanoassemblers/disassemblers will restore it to its original condition.

Please note: The above scenario is based on current common knowledge of the materials that are used in twenty-first century engines. I would be surprised if future advances in nanotechnology and materials science did not make current materials obsolete.

Know Your Place: Back to the Serapeum

As new arrivals on this planet, we begin life not knowing our place. From an early age until the end of our lives, we occasionally "take a walk on the wild side" and may still not know our place, but that is not because we have not sought it out or our fellow humans have not tried to show it to us. Knowing your bearings is critical for the successful navigation of space, whether it is off this planet or within the precise three-dimensional Cartesian coordinates inside the boundaries that define a work envelope on a CMM.

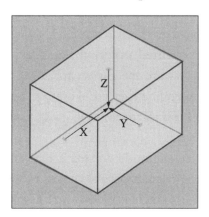

Figure 8.11. Example of a Cartesian coordinate system.

When we enter a factory as a worker, we are shown our place and taught how to behave and work within the parameters that define it. Within the work envelope of a CMM, the ruby ball tip of a touch probe is shown its place in three-dimensional space to within a fraction of a hair's width. That precise position is given as distances from an X0, Y0, Z0 absolute home point established by the machine, or a relative home point set by the operator. (See Figure 8.11.)

If we wish to build a perfect environment in which assemblers can successfully perform their job, there is no more perfect example than the boxes in the Serapeum. Let us see how this would work.

- The walls and underside of the lid define a Cartesian coordinate system within which the assemblers "know their place." (See Figure 8.11.)
- The bottom of the box is not a part of the coordinate system, as it will hold the materials that are used in the build. Therefore, exact precision is not required in that area.
- The outsides of the box and lid are not part of the coordinate system, so they do not need to be crafted with exact precision.
- The dimensions between the inside walls of the box can vary, as the coordinate system can be defined for each individual box by whatever dimension is finally achieved after critical attributes are met.
- The boxes are installed in a recess in the floor of the crypt so they do not move out of place when subject to vibration necessary to "unsettle" or mix the raw materials that otherwise would settle at the bottom.
- The boxes are installed in a space that can be filled with water that is heated or cooled to the required temperature.
- The temperature of the box is determined by the temperature at which the universal assemblers best perform their work, and the box is finished in situ at that temperature.

While discussing my observations of the precision found in the box I had tested using precision tools, I was challenged by one person to consider that my observations may not be accurate because the surface of these boxes may have a buildup of verdigris—a patina—that had accumulated over the centuries. I asked the person how thick the verdigris would be. He guessed that it might be 0.0005 inch (.0127 mm) thick, which is one-fifth the thickness of a human hair. My response was to ask him how flat a perfectly flat surface that had a coating of verdigris of .0005 inch thick would be then? I didn't get an answer.

Notwithstanding this person's motives behind the question, it raised a very important point that has now become more relevant. When considering how to communicate with the assemblers so they know where they are within a coordinate system, is it enough just to have passive, solid, flat surfaces of granite? Perhaps the walls and lid of these boxes were coated with an active material that could signal where it was located, and the universal assemblers could interpret those signals and know exactly where they were within that space at all times.

In his foreword to this book, Ahmed Adly describes the discovery of a previously unopened "sarcophagus:" *"Alessandro Barsanti explored the great pit of Zawiyet el-Aryan pyramid. Under 65 feet (20 m) of large blocks, he found a sealed giant granite sarcophagus. On removing the lid they found it was empty, except for black material that lined the side walls."*

Barsanti also noted that this so-called sarcophagus was "polished like a mirror," while on one occasion his companion, Alexandre Moret, noted, *"a splendid lid with four handles covers it, polished like a mirror and as carefully finished as an exquisite jewel."*

Today, Zawyet el Aryan is on property occupied by the Egyptian military, so it is off limits to tourists. Nevertheless, it is one of many sites that deserves thorough examination through the lens of advanced manufacturing. It is not, however, the only site that houses granite boxes with a black substance inside. It has been reported that at least one such box is located down the Osiris Shaft under the Sphinx causeway. I have not personally witnessed this but mention it in the hopes that it will inspire further investigation by professionals with a background in chemistry.

When discussing the building of a rocket engine, Drexler describes a system in which assemblers not only know their place but receive instructions on what to do. Similarly, when entering a factory to perform work, workers will be told where to go and what to do when they get there. Drexler describes a "seed" that contains a nanocomputer being located at the center of a base plate upon which the engine "grows." Stored in the nanocomputer are the plans for the engine. When the assembler needs to reorient itself and receive a new set of instructions, it plugs into the seed by sticking to it: "Since each assembler knows its location in the plan, it snags more assemblers only where more are needed. This forms a pattern less regular and more complex than that of any natural crystal. In the course of a few hours, the assembler scaffold grows to match the final shape of the planned rocket engine" (Drexler 1986, 61).

From Rocket Engines to Cutting Tools

Universal assemblers would not only be used for rocket or jet engines. Other products that would benefit from the advanced materials that these nanomachines produce could also be manufactured. Today, diamond is ground into small pieces that are sintered into metal and used for grinding. With nanotechnology, diamonds could be joined with other elements and be assembled into innumerable products, one of them perhaps being used to efficiently cut hard igneous rock.

Achieving Manufacturing Nirvana and the Return on Investment

To have just one box quarried, delivered, rough machined, semifinished, and finally finished in the crypts in the Serapeum would be enormously expensive. Three of the four precision granite manufacturers that I contacted failed to provide me with a quote for making one. Tru-Stone in Minnesota said they did not possess the equipment to produce one out of a single piece of granite but offered a plan where they would create the box using five individual pieces that could be bolted together after being delivered.

This has prompted some researchers to say that these boxes could not be made today. I disagree. They most certainly could be made today, but they would be very, very expensive. How expensive? I really don't know how many zeros to put in that number. Also, in order to spend those multiples of zeros, one would need a very good reason for it and the return on investment would be an important factor that would support the reason.

A very good reason could be that the Serapeum was created to serve as an advanced, state-of-the-art manufacturing facility. What was produced there brought value to the producers that would pay for their investment many times over. These products could be created in ways that we can only dream about and perform tasks that we are just now beginning to recognize as being able to exist.

But now as I find myself further out on the limb, I must return to address the skeptics who have accused me of fudging and faking measurements, without even considering that there may be valid reasons for the existence of precision and nonprecision that would save them the trouble of defaming a fellow researcher.

Once an area of an artifact is discovered to not be as precise as another area, the skeptic draws your attention only to the inaccuracies, as if that defines the artifact. With respect to this well-used logical fallacy—known as *post hoc, ergo propter hoc,* where a conclusion is reached based on one data set while other data sets are dismissed and ignored—there are two major points I would like to make. Every manufacturer understands them. It's not rocket science and can be learned by anyone who is willing to expand their understanding of a field within which they have no experience.

- High precision and low precision can and do coexist on the same manufactured item.
- The area on a manufactured item that is the most precise is more important to the customer than the area that is less precise.

For example, the granite surface plate upon which a CMM navigates is required to be calibrated periodically. This is performed by specialists like John Barta, who has decades of experience performing calibrations and resurfacing of inspection plates of all sizes. When observing this kind of specialist perform their work, if they strayed from the area that they were supposed to be calibrating and began to inspect the edges, the underside, or used a square to check the corners, one would be right to conclude that they were wasting time and did not know what they are doing.

And yet, we now have skeptics and debunkers traveling to Egypt to film in the Serapeum for the purpose of creating heavily monetized YouTube videos in which they accuse me, without evidence, of lying, fudging, and faking. There was no faking and fudging. I used inspection tools I was familiar with and had used successfully for many years. I applied these inspection tools to the inner surfaces of a granite box in the Serapeum and two photographs were taken (shown in figure 8.1 on page 185).

I can, and have, shown experienced toolmakers/machinists/ engineers these photographs, and the only explanation I give is where the artifact is located. Nothing more needs to be said.

Similarly, thousands of similarly trained specialists around the world have recently been riveted to learn on the UnchartedX channel

Figure 8.12. Tools of the trade. Are these fudged or faked?

the latest discoveries that have been made pertaining to the predynastic Egyptian vase discussed in the Introduction.

I challenge all skeptics to come forward to present their evidence and provide answers to the following questions:

- Which measurements do they object to?
- How were they fudged or faked?

EVIDENCE OF COMPUTER GUIDED MACHINES IN ANCIENT EGYPT

When it comes to crafting precise artifacts out of igneous rock, the observation that ancient Egyptians used machine technology was first proposed by Sir William Flinders Petrie in 1883. My article in the August 1984 edition of *Analog Science Fiction and Fact* expanded on Petrie's work, but in hindsight was timid. That article, "Advanced Machining in Ancient Egypt?" had a question mark at the end. Then, in *Lost Technologies of Ancient Egypt*, I presented additional evidence that prompted me to remove the question mark. Now, in 2023, through the efforts of other researchers, I can confidently place an exclamation mark where previously a question mark held sway! Yes, there *was* advanced machining in ancient Egypt!

I strongly advise the reader to access and read Mark Qvist's article "Abstractions Set in Granite" on his website, Unsigned.io. It is discussed in detail later in this chapter and I have also made it available on my website, Giza Power. Those who may think I did a fair job of describing ancient Egyptian machining will find that this research is light years ahead of what I was able to accomplish. I can only imagine what the next fifty years will reveal.

The Egyptian Vase Scanning Project

The inspection of Egyptian vases was initiated by Adam Young who from an early age had an interest in the ancient Egyptian culture and antiquities. This interest in his formative years quickened in 2010 when he and his wife, Emily, watched the "Pyramid Code" five-episode documentary series by Dr. Carmen Boulter (1954–2022), and he became impressed with a legendary figure who lived in a village called Nazlet El Samman near the Giza Plateau. Tall, gentlemanly, and charming, Abd'el Hakim Awyan (1926–2008) was blessed with a gift for relating native wisdom he had gathered over a long lifetime of guiding visitors around the pyramids and up the Nile, Hakim had gained quite a following of admirers around the world and spent some of his time visiting various countries giving lectures.

After Hakim's passing, his family continued his tradition of guiding visitors through Egypt, and his son, Yousef, a talented sculptor, now tours with Ben Van Kerkwyk of UnchartedX. Adam reached out to Hakim's family and joined one of their tours, which he remembers as a "life changing" experience.

I asked Adam what prompted him to build a collection of ancient Egyptian vases. He said, "from traveling with, and speaking to various Egyptologists and independent researchers, it was clear that the remnants of a vast and sophisticated body of knowledge were still present. Whatever happened in the distant past, there is not much physical evidence left, but we can see the results of precision engineering in granite and diorite stonework all throughout the world. We see it in large megalithic works such as the pyramids and other monumental structures, but we also can find it in the smallest works, such as granite vessels* and other objects."

Adam then actively searched for these vessels on the antiquities market. He remarked that it has become harder recently to find them, due to a greater interest now being shown than before. But, he notes, there may be more than a hundred thousand of these artifacts in private and public collections around the world.

In 2018, Adam joined my Lost Technologies of Ancient Egypt tour and became friends with my son, Alex. During that time they discussed taking some of Adam's vases and subjecting them to the same examination manufacturers use to inspect the quality of complex machined parts. Alex worked in the quality department at a defense contractor in

---

*Only one vessel is discussed here. It was purchased with the label "Red Amphora Jar" and hereinafter is referred to as "vase" and "object."

Indianapolis, so it was arranged for Adam to fly to Indy with his vases and subject them to traditional well-known and accepted mechanical methods for inspecting roundness, flatness, and concentricity.

I was a witness at that time and was quite surprised at the precision that was revealed when a quality inspector, with the vase mounted on a precision rotary table, and using an Interapid indicator, revealed a bore diameter that varied a mere 0.002 inch (0.050 mm). Other features, such as the top and bottom flat surfaces, the body diameter, and the lugs were similarly precisely round and concentric, to themselves and each other.

Adam and Alex's enthusiasm over these results cannot be questioned. To quote Adam, he said, "The results here speak to a fully-developed design and manufacturing system capable of mass production—using some of the hardest materials on Earth. These vessels are simply not possible to create by hand."

While I agree wholeheartedly with that statement, in the back of my mind was a certain degree of skepticism and concern about how far this could go in providing more proof of advanced machining in ancient Egypt. I was concerned about the item's provenance and was quite aware that skeptics would use this same concern to reject the item as not being authentic. At the same time, there would be arguments made that the vase was a forgery and manufactured using modern machine-tools.

Not daunted by such concerns, both Alex and Adam were inspired by these preliminary findings and made plans to expand the scope of their research to include an inspection method that was just being introduced into Danville Metal Stamping before I retired at the end of 2012, but one which Alex had received training in and performed on a daily basis in his job. That is structured light 3-D scanning. This technology, which is common today, acquires millions of data points on the surface of an artifact to create a point cloud out of which an STL file can be generated. ("An STL file describes a raw, unstructured triangulated surface by the unit normal and vertices (ordered by the right-hand rule) of the triangles using a three-dimensional Cartesian coordinate system." (Wikipedia). The STL file can then be analyzed in CAD software where geometric elements can be recognized and analyzed for size, form, and relationship to other elements.

Adam returned to New York with his vases, intent on taking them the next step. He said, "The quickest way to seeing the current narrative change (in my lifetime) is by embracing big data and AI. Hard data is irrefutable, but we need to centralize as much as we can to make any real statistical or analytical work possible. To that end, the goal is to scan as many ancient artifacts as possible. There is no other way in my opinion."

It was not long after Indianapolis before Adam had secured the services of Capture 3-D, a Zeiss company in South Windsor, Connecticut, who scanned a vase and provided him with the resulting STL file. Adam provided the file to Alex, who then collaborated with a coworker, Nick Sierra, a Purdue University graduate working in the metrology department of a composite specialty aerospace manufacturing company in Indianapolis, who performed an analysis in the PolyWorks CAD system.

After holding the results of their analysis for some time, Alex got in touch with Ben Van Kerkwyk of UnchartedX and described the project. Ben then invited Alex, Adam, and Nick on his podcast to present their findings. The comments section on Ben's YouTube channel became very active following the release of this video, with many commenters identifying themselves as machinists, toolmakers, and engineers with decades of experience claiming that the vase could not be made today. I had always thought that those who make this claim may be exaggerating, or not informed of the capabilities of modern engineers, but this flood of qualified opinions on the subject was impressive and persuasive.

Adam Young thought so too, and said, "It is heartening to see the sheer number of qualified engineers and scientists becoming interested in the subject. This is exactly what was lacking from the field of archaeology. This is how change will happen."

Prolific author and researcher Ralph Ellis emailed me his unreserved opinion of the vase. He writes, "One thing is for sure, Alex has proved this is not a recent forgery. No forger would go to this level of precision . . . !" (Email received on March 6, 2023)

Some, but not all, of my reservations about the artifact had been eliminated.

Particularly, the one feature of the vase that struck me was that its body closely matched the shape of an ellipsoid. In *Lost Technologies of Ancient Egypt*, I had identified the use of ellipses and ellipsoids in the design of statues, crowns, and temples in and around Luxor, commenting that in our history, credit for mathematically describing the ellipse was given to the Greek mathematician, Menaechmus, (380–320 BCE).

This vase, therefore, contained at least two attributes that I had previously pushed back in time and accredited to ancient Egypt: precision machining and the ellipsoid.

Following the interview with Ben, there were many requests from engineers to have the STL file available to the general public, so engineers all over the world can perform their own analysis and report their results. In considering this request, the team agreed to make the STL file available for download, arguing that data obtained from ancient artifacts should be "open source" because all inhabitants of the planet have a vested interest in the truth about their ancient ancestors. The desire was that full transparency should exist during this process, and "peer review" would take place without filters or censorship.

Figure 8.13. Egyptian vase overlayed with an ellipse.

When it comes to analyzing data, there are no better investigators than cryptographers and mathematicians. Alan Mathison Turing (1912–1954) is one of the most famous cryptographers in history. Turing worked at Bletchley Park during World War II, leading a team that was working on de-encrypting messages from what was known as "The Enigma Machine," which coordinated activities of the Nazi war machine. The results of his and his team's work is credited to have shortened the war and saved millions of lives.

Throughout my research in Egypt, I've always seen my role as a pointer. My books describe artifacts and point to features that I believe should be thoroughly analyzed using the most advanced technology and intelligence available. To the best of my ability I have used rudimentary tools I was familiar with to reinforce my argument. In the case of the predynastic Egyptian vase, this is the first time an Egyptian artifact has been subject to the most advanced methods for data collection, and now the most intelligent and knowledgeable scientific analysis capabilities have been used to analyze the results.

Mark Qvist, a cryptographer and mathematician, became intrigued by the challenge presented by the vase. He downloaded the STL file after it became available and set to work decoding the information that was collected from the outside surface of the vase. What he revealed is not just remarkable, it is revolutionary. He writes:

> When designing an object, be that functional or purely aesthetic, one can take a multitude of approaches. It is of course possible for the designer to simply place various features of an object intuitively, without any underlying rules, constraints or principles, but practically all great design, art and architecture follow sets of internally consistent principles.
>
> Various systems of design principles have been known since antiquity, and when beauty, completeness and harmony in form is experienced, it is often exactly because of the skillful application of such principles. From the preliminary analysis of the object, we deemed it warranted to investigate whether such a set of principles had been employed in its creation, and if so, to what extent.

An important data point that can be extracted through this approach is the degree to which different design patterns are locked together to form constraints. In design work, such constraints are commonly used to define various aspects of the individual features that make up the finished object.

If we can recreate the object from a relatively simple set of design primitives and constraints, which exhibit low degrees of interrelation, or if indeed no such relations even exist, and all features appear intuitively or randomly placed, this would indicate an object of a relatively low level of design complexity, or one that could have been made entirely from an intuitive process.

Contrarily, if we consistently see features defined by complex interrelations between different principles, this indicates a high level of design complexity, and that a higher level of abstraction was needed; not only in creating the design, but also in transferring it through the production method, to the finished object itself.

The Parametric Model

To explore what kinds of design principles were used in the creation of the object, we started out by measuring and mapping as many features of the object as possible, and looking for repeating patterns of placement and dimensioning, and repetition of mathematically significant ratios.

While we initially did not understand the underlying principles of what we found, it was clear that patterns and persistent ratios were present in overwhelming abundance throughout the object. Not only were they present, but they exhibited such a high degree of regularity, that we suspected them to be derived from well-defined mathematical formulae.

This led us to attempting something, that really should not have been possible, when dealing with a supposedly ancient artefact, made from granite, of all things: We decided to experimentally build a CAD model that would exclusively use mathematical concepts to dimension and place the features of the object, and use no tuning or arbitrary positional adjustments. All features should be placed and dimensioned by interrelation to each other.

We also limited the initial margin-of-error tolerance of the model to 75 μm, in terms of how well it should map to the features of the actual object. This margin-of-error tolerance was, amongst other criteria, used to discern whether an attempted modelling of a particular feature should be considered for inclusion in the model, or rejected.

I find it timely to stress how completely ludicrous this actually is. We are dealing with a stone vessel of supposed ancient origin, and are now proposing, that a purely mathematical CAD model, should somehow map to the actual object within a tolerance of less than 75 thousands of a millimeter.

Yet, I will let the results speak for themselves. Additionally, the CAD model, and all of its constituent equations, are available for verification, for anyone interested in doing so.

As Mark and his associates explored the design principle and underlying geometry and mathematics of the vase, a design familiar to most admirers of Egyptian art began to appear—the Flower of Life.

Figure 8.14. Egyptian vase overlayed with the Flower of Life.

Figure 8.15. The Ramses Statue overlayed with the Flower of Life illustrating correspondences with chin, mouth, nose, and eyes.

Thus appeared the third attribute in the vase that corresponded with other Egyptian artifacts. The design of the face of Ramses, crafted in symmetrical three-dimensional precision, quite elegantly accommodated the geometry of the Flower of Life, as presented in *Lost Technologies of Ancient Egypt.* However, its appearance, I will bet, if seen by a skeptic would cause a reaction akin to a silver crucifix being held up in front of a vampire. Repulsion. I can almost hear the howls of "New Age Pseudoscience" echoing in the hallowed halls of academia.

When I came to this part of Mark's article, rather than being repulsed, I became curious whether perhaps he knew of me and my work before he wrote his article. In the interest of science, I was hoping that he had not. As it turns out, after receiving a response to my inquiry by email, Mark informed me that he had no knowledge of what I had written before performing his analysis.

With three attributes being shared between the vase and other machined artifacts in Egypt, could I expect or hope for more? It seems like that would be enough, and I should be satisfied. For sure I was more than satisfied, but that didn't mean that there wasn't more. Also existing in Mark's decoding of the vase is the most basic and important revelation of all—the measurement standard that was used to design and manufacture the vase!

Everything we manufacture today is based on a measurement system. Technological advances in laser measurement resulted in a decision to establish base measurement to be the distance light traveled in a second in a vacuum. That would be 299,792,458 metres per second.

Surely the base measurement system used by the manufacturer of the Egyptian vases could not improve on that? Perhaps not, but would we be correct in assuming that they arrived at the same measurement system? Mark Qvist noted in his article: "We do, however, see an interesting *correlation* in this number. The approximation we have arrived at is just a mere 2.07μm (less than the scan accuracy) away from exactly matching the wavelength of an electromagnetic wave, with a frequency of 16 GHz, propagating in vacuum."

In summarizing his analysis of the vase, Mark made some stunning statements which in my previous research and analyses I only implied. That is the prehistoric use of computer aided design and computer aided manufacture (CAD/CAM). He writes, "Could an object like this simply have been a chance happening? A rare coincidence of random alignment? No. Proposing that would be completely magical and superstitious thinking. Maintaining absolute precision and consistency by chance, between all the interlinked systems present in the object, is —simply put—an impossibility."

Addressing the physical requirements to manufacture such an artifact, Mark does not mince his words or equivocate.

> As far as we know, no human beings, trained animals or naturally occurring phenomenae, modern or ancient, take mathematical formulae and equations as input, and produce lathe-operating motions as outputs.
>
> For all of the knowledge and insights we have accumulated over the ages, we know of exactly one, and **only** one *category* of things capable of such behavior: The kind of thing, that we refer to as a *Turing machine*. A device capable of taking input, holding state, performing operations on held states, according to pre-determined principles, and producing output.
>
> They come in many shapes and sizes, and can be constructed mechanically, electronically and even pneumatically or hydraulically. And you are most likely using one right now, to read this article.
>
> *We* call this class of device a *computer*, and no plausible way of representing, operating on, or manufacturing the design of this artefact exists, without having access to one such.

The research on Adam's vase has been a huge step forward in recovering our lost heritage. But it is not finished and much more needs to be done.

And more *has* been done. While a full analysis has not yet been performed, another scan of the Egyptian vases took place on April 12, 2023. This time it, along with three others, were X-ray CT scanned at Zeiss Quality Excellence Center in Wixom, Michigan. Eric Wilson, Adam Young, and I met in Wixom the night before the scan and signed in at Zeiss's impressive state-of-the-art facility the following morning.

This scan was performed because the structured light scan did not produce any data from the inside of the vases, and there was a keen interest in knowing the geometries and precision of the inside compared to that of the outside. The file that was produced by the scan was 80 gigabytes. There has been a cursory examination of the data, but more work needs to be done before releasing any official results. Needless to

Figure 8.16. Adam Young attending the CT scan of his Egyptian vases.

Figure 8.17. Metrologists Nick Sierra and Alex Dunn.

say, what has been revealed has left us scratching our heads.

From my perspective, this research program has proven to me that the vase's origins are clearly Egyptian. However, I understand that not everyone may agree with me and another level of investigation will need to take place. But we now have convincing hard data that should encourage the Egyptians to subject similar artifacts with known accepted provenance to examination using similar or more advanced technology and knowledge. Considering how young Egyptians are now accepting this new understanding about their ancestors, I anticipate that in the future they will insist on this taking place.

What will be revealed by a systematic study of a thousand similar artifacts of known provenance? Will we eventually view these caches of vases as time capsules collectively encoded with more of their creator's knowledge? With each individual vase holding a different piece of information that fits within a larger data set?

Considering the period of time in which I have witnessed huge advances in manufacturing technologies, the ten years that have passed since I retired have resulted in so many improvements, it is inspiring. Comparing a modern manufacturing plant with those I worked in 60 years ago, the difference is like night and day. What will the next 60 years look like? With the evidence presented here of the level of manufacturing that was lost in prehistory, it is not hard to imagine further developments in the future that will result in adherence to similar standards of precision.

When I consider the capabilities that have been observed by Unidentified Aerial Phenomena (UAPs) and ponder on the level of manufacturing excellence that would be required to build these machines, is it too far out on the limb to suggest that micro-machining was taking place in a vacuum and was free of gravitational and magnetic forces? Of course it is! What a ridiculous idea! Nonetheless, I remember a time in my career when a machine of high precision was installed in the plant, and the manufacturer's installation specifications included how the axes of the machine should be aligned to the Earth's magnetic field.

What quality improvements can be made if shafts that have no weight rotate inside bearings that use no oil?

It is a commonly held view that UAPs are the product of a society that has manufacturing capabilities that exceed our own. Speculation about how many years of development it took to achieve their capabilities usually exceed by far the number of years it has taken our civilization to get where we are. The UAPs are teasing us with their capabilities and are finally getting everybody's attention. But they are elusive little buggers, and we cannot hope, with our current technology, to grab one and learn what makes them tick.

Similarly, a pyramid electron harvester would be astronomically expensive to build, but, as we will see in the next chapter, the investment would be quickly paid off and continue to provide value for an untold number of years.

The question, "Where are the machines?" is consequential. The answer to that question right now seems to be beyond imagination. That is, "there are some that are so astonishingly large and others that are so inconceivably small, you wouldn't recognize them."

As Mark concludes in his analysis,

> Only by open, critical and honest scientific inquiry, working from first principles anchored in concrete and verifiable data, can the deepest and most obscured echoes of our shared past be brought into the light of day once more.
>
> We must allow the observable facts, the givens that we have, to speak without prior interpretation, even if the conclusions risk temporarily upsetting a prior understanding.
>
> In the scientific endeavor, all theories are provisional, and must yield when a simpler one, that better fits the evidence, can be constructed. If this fundamental tenet ceases to hold primacy, science ceases to be science, and instead degrades into dogma (Qvist 2023).

And as Alex Dunn said in his interview with Ben Van Kerwyk, "Data does not have an opinion."

NINE
THE ULTIMATE OLD
AND NEW GREEN
ENERGY MACHINE

Egyptian scientist Gamal Elfouly and I began our studies of the Great Pyramid at around the same time—the late seventies. Without knowing or communicating with each other, we both concluded that the Great Pyramid was a machine, the purpose of which was to supply energy to the civilization that built it. While our models explaining how this was accomplished vary somewhat, our fundamental recognition of reality is the same: The pyramids are machines.

Gamal's assertion on this point could not be stated any better:

As human beings "The Great Pyramid is a tomb" theory is an insult to our intelligence, especially after finding its precision and workmanship cannot be duplicated today. This means that those who built it were more technologically advanced than us in this regard, and possessed a technology unknown to us that enabled them to do this miracle in stone. That's the reality of the matter whether we like it or not (Elfouly 2012, 35).

BUILDING PYRAMID ELECTRON HARVESTERS

Finance

According to the US Department of Energy, the cost of building a nuclear power plant ranges from $14 billion to $30 billion. For a new fossil fuel power plant, costs are becoming prohibitive. The 582-megawatt Kemper Power Station, in Kemper County, Mississippi, was designed with carbon capture and sequestration (CCS) technology and was supposed to be the largest coal power station in the United States. Its original cost estimate was $2.4 billion. Construction began in 2010 but in 2017, after costs tripled to an astronomical $7.5 billion, the plans for using coal were scrapped and changed to using a more eco-friendly natural gas. Outside the United States, the cost would be lower, depending on labor costs and environmental regulations.

What would the cost be to build a pyramid solid-state electron harvester the size of the Great Pyramid? Fouad Kamal, who has a background in infrastructure financing for large-scale infrastructure, particularly power plants, emailed me with his estimate of what the costs would be to build the Giza power electron harvester:

Hi Chris,

I came across your videos whilst researching Egypt and have been astounded to learn of ancient lost high technology, and also viewing the Giza complex as a power plant.

Anyway, the purpose of my message: I have a background in infrastructure financing for large scale infrastructure projects (especially power plants). I tried to estimate the total construction costs (modern day) for the Giza pyramid complex, and apply modern day financial returns analysis.

The objective was to reverse estimate how much Megawatts power (installed capacity) these Giza power plants would have been producing. The results of my analysis were simply incredible.

The numbers suggest it may have been a colossal plant providing energy to millions of people. Or perhaps a mega central power plant providing energy to all sorts of substations (all with those mysterious 70 ton granite boxes with precision shafts) which then passed the power to local areas. Same thing effectively.

A typical, standard coal/gas power plant today will cost a ballpark $1m per megawatts capacity to construct. (Land costs etc excluded). As a ballpark, 1 Megawatt capacity will provide power to around 150–200 homes, as well as industrial/commercial usage, for a location with an above average industrial capacity.

So a medium sized 300MW coal/gas plant in Egypt today would cost around $300m, and provide power to 150,000 urban residents, along with industrial/commercial use. I worked on arranging debt financing a few years ago for a European company to build a gas fired power plant in Egypt fitting these parameters.

So I asked myself, leaving aside our lack of knowledge as to precisely how and when it was done, how much would it cost today to build the Giza complex? And how much energy would such a power plant produce and for how many people?

Quarrying first: You would have to pay for 2.3 million stone blocks (average 3 tons) to be cut, all different sizes/shapes. Let's assume the quarry is publicly owned so the material (granite, sandstone etc) is free.

To cut a granite slab (wholesale) for a kitchen today (2.5m by 1m by 15cm) would cost around $500. It would be ridiculous to assume you could get a 3 ton block cut to unique custom shapes for only $2000 a piece, but let's go with that figure. This covers the cost of labour at the quarry, as well as paying for all the cutting/lifting machines there.

If you mistakenly assume you are paying zero for using the cutting/lifting machines, effectively you are just passing those costs elsewhere, because they themselves would need to be paid for. Let's assume there is a govt quarrying department, and one arm of the govt is paying this quarrying department just for its costs.

So the 2.3m blocks are $5bn for the great pyramid alone. Around $12.5bn when you include the 2 other main pyramids and the smaller ones.

I've assumed the second pyramid is the same size as the great pyramid, and the others all combined (including eg the 3 smaller pyramids next to the great pyramid) add up to another half unit.

Then the transport aspect. Today, SUVs are transported on large trucks that carry 8 vehicles on a double decker structure. The typical cost to transport an SUV in the United States is around $1000. So let's go with a $1000/block transport cost. Even this is very low, because you would need to drive a lot slower with precision stone than with an SUV (which itself has suspension), in order to prevent cracks. And this is assuming you have a modern world class motorway in megalithic Egypt in place already.

So the great pyramid blocks transportation cost would be around $2.5bn, and around $6bn for all the Giza pyramids.

Now for assembly on site. It would be ridiculous to assume a modern cost of $1000/block for receiving, unloading, sorting, queuing, lifting and placing 2.3m blocks in a precise sequence. There are all sorts of costs excluded here, such as blasting and flattening the bedrock underneath. But going with this figure provides a $6bn assembly cost for the Giza complex.

Note: I have excluded the enormous costs running into hundreds of $millions to architect/design a unique construction of unprecedented size with millions of interlocking blocks, and built to all sorts of mathematical specifications (eg pi/phi ratios, world circumference etc), and all sorts of other specs we don't understand today.

So the ballpark total cost of building the Giza power plant complex would add up to around $25bn. This is $12.5bn for the stones, $6bn for transport, and $6bn for assembly.

And there are all sorts of modern costs (eg insurance) excluded here. I've also assumed these costs are paid in cash, as opposed to getting debt financing from Bank of MegaEgypt and paying interest costs.

With modern metrics, this $25bn would provide around 25,000 MW (25GW) installed capacity. This would provide constant energy to around 10 million people today at Singapore's super high per capita usage. They would have fridges, freezers, TVs, lighting, microwave ovens, computers, washing machines, etc, as well as industrial/commercial usage.

Singapore, a small, densely populated, urbanized and highly industrialized country with extraordinarily intense per capita energy usage, has 12.5GW installed capacity for its approx. 5m population. (Energy Market Authority n.d.)

The entire UK electricity installed capacity is 75GW, with a 67m population. The 25GW installed capacity estimate of the Giza pyramid complex would therefore provide around 33% of the entire UK's power capacity! This equates to electricity supply for over 20 million people in the UK, along with industrial/commercial usage. (Wikipedia 2011)

If the whole Giza complex was a power plant, obviously they wouldn't have gone to such lengths in order to power a couple of kettles. That would make no sense for a rational project.

Viewing the Giza complex power plant from a financial analysis perspective will invariably produce a gigantic MW capacity. You could chop my assumptions in all sorts of directions (eg lowering transport costs) and you would still have an inconceivable number.

Even if you went so far as to chop my assumptions by 99%, this would still be enough energy for over 200,000 people in the UK today at full regular usage.

I would need a broader understanding of the entire pyramid network across Egypt, as well as all those shafts with the granite boxes, to estimate the total energy usage in the country.

However, taking this financial analysis approach is highly relevant and I hope others can start assessing the Pyramid power systems worldwide from this perspective as it will provide new insights.

The basic metrics of financial analysis for energy infrastructure projects are the same for all time. It would apply equally to a windmill in medieval Europe, for example. It would have had a cost to build, and a productivity assumption that would make the cost worthwhile.

I hope this helps.

Yours,

Fouad

Kampower Estimate Metric

(Fouad Kamal, email to author, March 27, 2021, 12:37 pm)

If we take Mr. Kamal's estimated cost of 25 billion dollars for an output of 25 gigawatts of power, providing constant energy to around 10–15 million people a day, a rough estimate can be made of the return on investment (ROI) in order to financially justify building the Pyramid Electron Harvester. I would assume that this estimate would be subject to change as time progressed and lessons were learned, but more importantly, there is much more to learn before a fully accurate analysis can be made. It would seem, though, that if there was financial and political will behind the research and development of such an endeavor, as there was in the early '60s when President Kennedy inspired the nation to support the mission to the moon, it could be achieved in a shorter time frame than the 80 years promised for nuclear fusion. Also, as Ashraf El Sherbini pointed out, it would require organizational power with money and connections, using top specialists in all the relevant fields to even arrive at a decision as to whether it was feasible to do it.

Power generation costs vary from country to country and region to region. These costs drive our electricity costs, which also vary significantly depending on where you live. It is not within the scope of this book to provide a detailed analysis of what the ROI would be for each individual power generation source but to provide an overall awareness of the comparison between two methods of achieving the same result. They both start with the extraction of rock from the Earth. The first rock I will discuss is coal.

COAL-FIRED POWER PLANTS

When reading the following, I don't want the reader to get the impression that I am against coal. How could I be against a material that all my life has kept me warm in the winter, cool in the summer, and fully employed for 52 years? A material that allows me to do what I am doing now—sitting in front of a computer typing this manuscript. Even in Manchester, England, when walking through the smog, which left a black ring on the inside of my shirt collar, I had no hostility toward coal and on a cold night looked forward to getting home and throwing a lump of it on the fire.

I remember waiting in excitement and anticipation for the day I reached the age of eight. This was the day that I was allowed to start the fire that would heat the house and provide it with warmth and hot water. Newspapers, kindling, and coal were carefully arranged on the grate—and I got to play with matches!

No. I love coal for the comfort it has given me. At the same time, I respect and admire our forebears and what they worked tirelessly to accomplish to make life more pleasant for following generations. Without their hard work, none of what we have would be possible. While growing up, I had an uncle who delivered coal to people's houses. My neighbor across the street was a coal miner who died of black lung disease. There were reports of mining accidents and horrible deaths, both inside and outside the mines. I learned about the role in the mines of canaries and Davy lamps that would alert miners to get out because of deadly gasses. I also learned about early mining when boys and girls, some as young as six, worked in the mines alongside pit ponies in order to provide for their families. I also learned about waste management when, on October 21, 1966, a slag heap piled on a hillside overlooking Aberfan village in Wales collapsed and took the lives of 116 children and 28 adults.

I didn't work in the mines, so I didn't get the opportunity to see miners at work. However, I did get to admire their spirit and talent when they left the mines: the Welsh miners who, when exiting the mine, raised their voices in song, and their contribution to cultural activities, like the famous Welsh male voice choirs and works bands that competed with their counterparts throughout Britain. Also, I cannot help admiring the resilience of nature as flowers and trees flourished in the slag heaps piled up from mining operations. And miners did not just produce coal and beautiful music. Their children have inspired and entertained millions with their talent. Singers like Tom Jones and Loretta Lynn, the famous actor Richard Burton, and many more. All the while, I still happily and gratefully burnt the coal, either directly or by proxy when paying my electric bill.

However, science and technology evolve, and existing technologies and methods are discarded or phased out after new and better ways of achieving the same results with greater efficiency and less cost are adopted, which brings me back to the discussion about new physics, new science, and new technology for accessing electricity. Just like when burning coal, we don't create electricity; it already exists. There are various methods used for harnessing it, and each method has its own costs and benefits. My intention is not to provide an in-depth analysis of each, as, unlike Fouad Kamal, I am not an expert. In that same vein, the information I have learned about coal, while provided by experts, probably strays from the real truth with respect to accuracy, and I recognize that others with more expertise in the subject might provide more accurate information.

But what I have learned through accessing information available on the subject of coal has raised my awareness about the sheer weight of coal that is mined every day and the primary and ancillary operations needed to mine it, transport it, deliver it, process it, and feed it into furnaces that burn it to create steam.

Volume of Coal

In 2021, the weight of coal used to create electricity in the US was 461,187,000 metric tons. This is based on a total output of 828 billion kilowatt hours using 1.13 pounds of coal to produce 1 kilowatt hour of electricity (USEIA n.d.).

The Great Pyramid is estimated to weigh around six million tons. Therefore, the weight of the coal extracted to create electricity in one year in the US is enough to create an enormous pyramid 76 times the weight of the Great Pyramid at Giza.

Like many others, when I visit Egypt and stand in front of the Great Pyramid, I am awestruck by its sheer size. A precision, artificial mountain covering thirteen acres that commands attention and has the power to distract visitors from what surrounds it. Now I am awestruck to learn that the coal mined in one year in the United States, if stacked beside the Great Pyramid, would be 76 times its weight and cover almost a thousand acres. Certainly, it would not be cut and stacked with the same precision, but both the coal and the

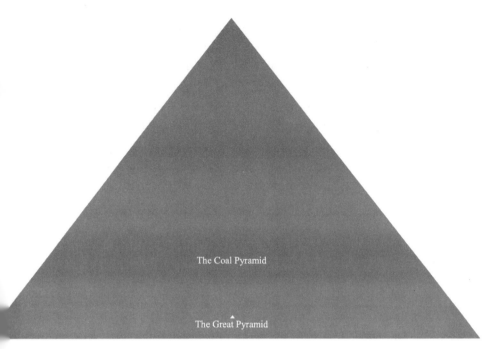

Figure 9.1. The size of the Great Pyramid (shown with small white triangle) compared to the amount of coal extracted in one year to fuel power plants.

rock that was used for the pyramid came from the same source—our planet, upon which we merely scratch the surface in order to survive. A planet that gives us life and enjoyment while occasionally shaking things up and making adjustments to its topography. Adjustments that sometimes remind us we cannot compete with the Earth in our ability to change its landscape, regardless of how aggressive we become.

The US Energy Information Administration reports that the number of workers employed in the coal mining industry in the US has shrunk from just over 90,000 employees in 2011 to just below 40,000 in 2021. That's a lot of workers who have demonstrated the ability to extract millions of tons of rock. Taking this into consideration, the prospect of building a new Great Pyramid becomes less daunting.

LOCATION OF ELECTRON HARVESTER

Ideally, if we decided to start construction on a new Great Pyramid in the United States, we should learn from the ancient Egyptians and locate it on limestone bedrock where there is sufficient material surrounding it to extract and build the bulk of the structure. Perhaps one location to be studied would be near what has been called the limestone capital of the world: Bedford, Indiana. The quarries around Bedford have provided limestone for many famous buildings, such as the Empire State Building and the Waldorf Astoria in New York, the Pentagon, and also, most recently, the beautifully crafted Palladium at the Center for the Performing Arts in Carmel, Indiana.

Of interest, in the late 1970s, the town of Bedford received government funding to build a pyramid, with the intention of creating a tourist attraction, and soon had a pyramid under construction near a giant limestone quarry near Needmore, Indiana, about nine miles away. Unfortunately for the project, in 1981, Senator William Proxmire, who had created the Golden Fleece Award with the intention of sniffing

Figure 9.2. The Palladium in Carmel, Indiana.

out what is known as "pork" spending, got wind of the expense, and the pyramid park became that year's unlucky recipient of the award. The project received no more federal funds, and work on the pyramid ground to a halt (See "Cursed Pyramid Ruins" on the Roadside America website) (Roadside America n.d.).

It would all depend on the geology of the area, but it's interesting to note that Bedford is located 200 miles from the New Madrid Fault, Missouri, which is recorded to have produced two major earthquakes that changed the topography of the area: on January 23, 1812, a magnitude 7.0–7.8 earthquake, and on February 7, 1812, a 7.4–8.0 earthquake. New Madrid is located in the Mississippi Valley in an area known as the "bootheel" of Missouri. It lies around 164 miles south of St. Louis. Observations during and after the earthquake were:

- **Sand Boils:** The world's largest sand boil was created by the New Madrid earthquake. It is 1.4 miles long and 136 acres in extent, located in the bootheel of Missouri, about eight miles west of Hayti, Missouri. Locals call it The Beach. Other, much smaller, sand boils are found throughout the area.
- **Seismic Tar Balls:** Tar balls ranging from small pellets up to golf ball size are found in sand boils and fissures. They are petroleum that has been solidified, or petroliferous nodules.
- **Earthquake Lights:** Lights flashed from the ground, caused by quartz crystals being squeezed. The phenomenon is called seismoluminescence.
- **Warm Water:** Water thrown up by an earthquake was lukewarm. It is speculated that the shaking caused the water to heat up and/or quartz light heated the water.
- **Earthquake Smog:** The skies turned dark during the earthquakes, so dark that lighted lamps didn't help. The air smelled bad, and it was hard to breathe. It is speculated that this was smog containing dust particles caused by the eruption of warm water into cold air.
- **Loud Thunder:** Sounds of distant thunder and loud explosions accompanied the earthquakes.

- **Animal Warnings**: People reported strange behavior by animals before the earthquakes. They were nervous and excited. Domestic animals became wild, and wild animals became tame. Snakes came out of the ground from hibernation. Flocks of ducks and geese landed near people (New Madrid n.d.).

BUILDING THE ELECTRON HARVESTER

To get a rough idea of how long it would take just to quarry and deliver to the site the limestone upon which the pyramid electron harvester would be built, in 1978, author Richard Noone requested a time study from Merle Booker, the technical director of the Indiana Limestone Institute of America, of what it would take to quarry, fabricate, and ship enough limestone to duplicate the Great Pyramid. Perhaps because of the natural and professional appreciation that modern limestone workers have for those who performed the same work in ancient times, Booker performed a detailed analysis and provided Noone with his response as follows:

Dear Mr. Noone:

As a follow-up to our letter, dated June 26, 1978, we have various Indiana limestone industry members working out realistic quarrying and fabricating time studies for all of the necessary blocks that will be required to duplicate another great pyramid, like the one built in Egypt centuries ago.

Utilizing the entire Indiana limestone industrial facilities as they now stand, and figuring on tripling present average production, it would take approximately 27 years to quarry, fabricate and ship the total requirements. The quantities of various types of blocks were based on a 755' 9" × 755' 9" base figure for the pyramid and the side slopes of 51° 51' 14.3". Also based on providing a king's chamber, queen's chamber, grand gallery, ante-chamber and passageways on the interior.

All exterior and interior blocks were figured on the basis of 12' 0" × 8' 0" × 5' 0" size. Approximately 264,216 rectangular core blocks will be required. Plus 12,723 exterior sloped blocks with very precise joint surfaces. The total quantity of blocks required would equal 131,467,940 cubic feet of quarried finished stone.

Sincerely,

Merle B. Booker

Technical Director

(Noone 1986, 105)

Even after lengthy research and reaching agreement that there would be benefits to building a pyramid electron harvester, the length of time it would take to build one would not be just 10 years or 100 years, as has been suggested by some. If things went well, the time frame to build another Great Pyramid would be somewhere in between, depending on the size of the quarry, the number of quarry workers, and the size and type of machines that are used. If we look at the accomplishment of one man, Edward Leedskalnin, with proper organization and without unnecessary interference, I would estimate that it would take less time to build it than it took to make the decision to build it.

"I know the secrets of how the pyramids were built!"

While many theorists have made the claim that they *know* how the pyramids were built, I'm not one of them. However, there is one person who left evidence when he died that he had a valid reason for making this claim, because he left standing in Homestead, Florida, an impressive legacy that people flock to daily to marvel at what he accomplished. Evidence of this Latvian immigrant's time on the Earth is not defined by his simple grave marker in Miami Memorial Park Cemetery; Leedskalnin's real monument can be found 30 miles south on Route 1 at a popular tourist attraction now known as Coral Castle. There you will see 1,100 tons of quarried and standing coral bedrock, with some pieces weighing up to 30 tons, providing a testament to the fact that he had a good reason for making his claim. He possessed an extraordinary knowledge that followed him to his grave.

Unfortunately, he did not share his secret and refused to allow people to watch him working. There is a full chapter on Edward Leedskalnin's

Figure 9.3. Edward Leedskalnin's grave marker in
Miami Memorial Park Cemetery.

Figure 9.4. Coral Castle, Leedskalnin's true monument,
in Homestead, Florida.

Coral Castle in *The Giza Power Plant*. However, I believe we don't fully "know" Edward Leedskalnin, and there is still much more to learn. Nonetheless, when it comes to roughly estimating the time and number of workers it would take to build the Great Pyramid, what he accomplished may give us a rough idea.

> If we assume that Leedskalnin and the ancient pyramid builders were using similar techniques, we must reevaluate the requirements for the man-hours necessary to construct the Great Pyramid. Estimates provided by Egyptologists for the number of workers that built the Great Pyramid range between 20,000 and 100,000. Based on the abilities of this one man, quarrying and erecting a total of 1,100 tons of rock over a time span of twenty-eight years, the 5,273,834 tons of stone built into the Great Pyramid could have been quarried and put in place by only 4,794 workers. If we figure in the efficiencies to be gained from working in a team and the division of labor, we can reduce the number of workers and/or shorten the time needed to complete the task (Dunn 1998, 112).

When Sir William Flinders Petrie measured the casing stones of the Great Pyramid, he found that the precision of the surfaces were within .010 inch (.254 mm) over 75 inches (1.905 m)—half the thickness of a thumbnail. He made the observation that it was: "Accuracy equal to most modern opticians' straight-edges of such a length" (Petrie 1883, 13). It has been stated that Petrie said the casing stones were finished to optical precision. Considering what he wrote, I suspect he might correct that interpretation. Optical precision generally refers to precise surfaces controlling the behavior of light through reflection or refraction that is dependent on wavelength precision. Petrie was referring to something else entirely.

When I first measured the flatness of the inner surface of the box inside the second largest pyramid on the Giza Plateau (Khafre's), I exclaimed, "Space-age precision." I think it is important to explain what my exclamation was founded on. When I first discovered the precision in Khafre's pyramid, I was shocked. When I said "space-age precision," it was a true statement founded on my background and experience working in aerospace manufacturing. This does not mean that precise flat surfaces were not required in products before we entered the space age. Or that the ancient Egyptians were actively involved with space exploration. What it meant was that if we go back in time and identify a period when precision was necessary on flat surfaces, such as when Johannes Gutenberg developed his flat platen printing press around 1400 AD, his press would not compare to the boxes in the Serapeum or the box in Khafre's pyramid in terms of the precision required and achieved. I can say that from experience, because I was taught during my apprenticeship in England how to file, lap, and scrape a flat platen to match the precision of a surface plate, which is used in manufacturing to inspect precision components that are used for a wide variety of products.

Possessing practical experience in creating precise flat surfaces, which I didn't achieve by watching a video or reading an internet web page, my reaction to what I had discovered was obviously heavily influenced by comparing what I was seeing in Egypt with my own efforts—which were paltry in comparison. Other engineers and craftspeople can form their own opinions, but of those with whom I have discussed this matter, none have objected to or questioned mine.

My reason for sharing the above information is because it has relevance when discussing the building of a new pyramid electron harvester with the same features as the Great Pyramid in Egypt. The relevant factors are a necessity for achieving accurate geometry and precision where it is necessary and not trying to achieve it where it is not. Both conditions of manufacturing have their own costs, with high precision obviously costing more in terms of machines, time, and skilled workers, who are typically paid more for their labor. Knowing what is required of skilled workers, in *The Giza Power Plant* I discuss the time I visited a quarry worker to learn his perspective.

Between 1969 and 1984 I lived sixty miles away from Bedford. One day I took an easy and pleasant drive through the picturesque southern Indiana countryside, which was ablaze with fall foliage, to talk to Tom Adams, who at that time worked at one of the quarries. Adams worked in the shop, cutting and dressing the stone, and the accuracy he was required to maintain in his work is not as stringent as for those who work with machine tools. Any craftsperson in a tool shop or machine shop can tell you exactly the tolerances they are working to. I asked Adams about the tolerances they work to in the quarries. He answered, "Pretty close." I asked, "How close is pretty close?" He responded, "Oh, about a quarter of an inch." (6.35 mm) Adams was astounded to hear that the limestone in the Great Pyramid was cut to .010 inch tolerance. His response regarding the abilities of the pyramid builders confirmed my belief that, contrary to what we have been taught, the pyramid builders were not primitive workers of stone.

It was clear to me that modern quarrymen and the ancient pyramid builders were not using the same set of guidelines or standards. They were both cutting and dressing stone for the erection of a building, but the ancient Egyptians somehow found it necessary to maintain tolerances that were a mere four percent of modern requirements. Two questions sprang from this revelation. Why did the ancient pyramid builders find it necessary to hold such close tolerances? And how were they able to consistently achieve them? (Dunn 1998, 55–56)

With respect to the requirements and steps to take (except for how the blocks were moved and lifted, the builders of the Great Pyramid left us with guidance on how it was built—the Great Pyramid itself. It has been suggested by some that the Great Pyramid is a "time capsule" that was built with the intention of providing knowledge to future generations. While I might disagree with the idea that it was the sole "intention" of the ancient Egyptians to pass along the knowledge they had gained, structures of all kinds that were built in the ancient past certainly do provide valuable information and give us insights about the people who built them.

The most important part of a building is its foundation. As implied in the philosophy of René Descartes, whether it is the structure you build in your mind or the structure you build physically, if the foundation is not strong and true, neither will be the structure. Regarding the preparation of the foundation upon which the electron harvester will be placed, what we know about the Great Pyramid is that the perimeter of its base was level to within .875 inch (2.22 cm) over 13 acres:

> The level of accuracy in the base of the Great Pyramid is astounding, and is not demanded, or even expected, by building codes which existed in 1974. Civil engineer Roland Dove, of Roland P. Dove & Associates, Martinsville, Indiana, explained that .02 inch (.508 mm) per foot was acceptable in modern building foundations. When I informed him of the minute variation in the foundation of the Great Pyramid, he expressed disbelief and agreed with me that in this particular phase of construction the builders of the pyramid exhibited a state of the art that would be considered advanced by modern standards.
>
> In *Pyramid Odyssey*, William Fix stated that the most accurate survey of the base of the Great Pyramid showed it as 3023.13 feet (921.45 m) around the perimeter, with the average of the sides being 755.78 feet (230.36 m). If the alignment of this structure was governed by the building standards that existed in 1974, then one side of the Great Pyramid would be allowed a variation of 15.115 inches (38.39 cm) (Dunn 1998, 56–57).

What follows is a rough estimate of what steps may be taken to build the pyramid after a site is selected and approved for construction, the design work is completed, and a contractor is selected to oversee the project. With the project's completion deadline in mind, the main contractor will hire subcontractors working in mission-critical industries who separately and in concert with each other will:

- Estimate the number of workers required to meet the deadline and hire a sufficient number with the necessary qualifications that are needed. Quarries that have high-quality limestone needed for casing stones and interior chambers may decide to expand their operations and install precision machines to cut limestone with greater precision. If not, a separate facility closer to the build site may be built to receive the rough quarried stone and cut it to the specifications provided by the architect. Their machines should also be able to cut granite with precision and ideally will be located near the building site or elsewhere.

- Provide training to quarry workers (like Tom Adams) so that they are able to operate the machines and produce limestone and granite pieces with greater accuracy than what was required of them previously.

- In parallel with the above, contractors will prepare the base upon which the electron harvester will be assembled according to specifications. (Note: Specifications of .020 inch per linear foot accuracy will not be acceptable.)

- Design and build special machinery to accurately bore into the bedrock to create a rectangular passage and subterranean chamber to house pulse-generation equipment.

- Excavate pits in the bedrock near the electron harvester that will house the lower portion of a large saw. See "Under the Shadow of Egyptian Megamachines" (Dunn 2010, 247).

- Build a maintenance facility that will repair and/or replace cutting segments that are affixed to the outside of the saw. Also build administration building, canteen, restrooms etc.

Casing Stone Specifications

To achieve the geometry and precision required in the pyramid—and with an understanding that there is always the risk of an accumulation of error when joining many pieces together—as each layer of the pyramid is built, the casing stones should be the first blocks to be delivered to the site and put in place. They will be surveyed to ensure they are in the exact location they are meant to be according to design specifications, which will include orientation to a precise direction.

For instance, the Great Pyramid is known to be oriented to true north within three minutes of a degree. It is also known that each side of the Great Pyramid has two surfaces, which means that the distance from corners east to west on the north side is greater than the distance from the center of the pyramid on the east side to the center of the pyramid on the west side. (See Figure 9.5.) Therefore, the outer surface of the casing stone must be cut on a compound angle. This means that when viewing a casing stone along its length, with its length defined as the X-axis, its width defined as the Y-axis, and its height defined as the Z-axis, viewing it from north to south on the east side of the pyramid, the pyramid angle of 51° 51' 14.3" is seen on the left side. When viewed vertically from the top, the width of the north side of the casing stone is greater than the width of the south side of the casing stone. (See Figure 9.6.)

State-of-the-art tools that were used by the ancient Egyptians to accomplish this precise construction have not survived the millennia. State-of-the-art tools available today to build a pyramid electron harvester to the same specifications found in the Great Pyramid are powerful and efficient and when applied can provide quality control and assurance that the design specifications are met or exceeded.

For data acquisition and analysis, each casing stone and interior wall, floor, and ceiling block can be scanned by a laser, and the information gathered can be imported into a computer where it is placed within a model. Existing will be an efficient orchestrated, real-time, and continuous transfer of data between the architects, the building site inspectors, the data analysts, and the mill that will receive instructions if changes are needed to any parameters to machine the blocks.

The dimensions of the X- and Y-axis of the casing stones may vary depending on the size of block that is delivered by the quarry to the mill, but the casing stone that defines the center of the distance of one side must be cut to assure that the surface that touches the next block face (defined by the Y-axis and Z-axis) is dead center. Think of it as a giant Tetris puzzle, where players can make changes to blocks that are not quite the right size. Also, imagine the number of men and women around the world who have the qualifications to play this game. They

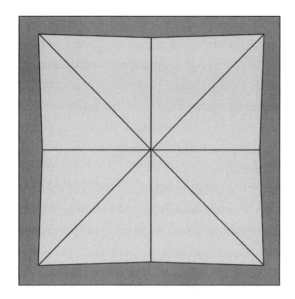

Figure 9.5. Top view of Great Pyramid.

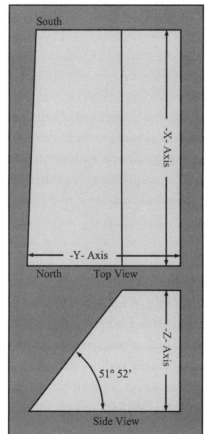

Figure 9.6. Geometry of a casing stone.

could work from home, download the scanned data from a block as it leaves the mill, place it within their computer model, and, if it fits virtually, authorize its installation. After the block is installed, they can then download the data acquired from the build site to verify its correct position. This process can be repeated for every casing stone that is installed.

Sequence for Assembling Blocks

1. Assign and affix a tracking serial number on every block that leaves the quarry before finish work is performed on them.
2. Upload serial numbers and dimensions of the block to a central database.
3. Install casing stones and, as the pyramid rises virtually and physically, install precision blocks required to build its chambers, passages, and shafts.
4. Install core masonry blocks. They can be of varying shapes and sizes that are cut efficiently on site using the giant saws. The tops of core masonry blocks positioned next to a casing stone will need to be cut on the same plane as the casing stones to provide a surface to accurately position another layer of casing stones. Perhaps, also, they should be made from a higher quality of limestone than those that do not have a casing stone on top of them. The same considerations would apply to the core masonry blocks touching and supporting the precise blocks that create the interior architecture.

Core masonry blocks that are installed between the outside casing stone blocks and interior architecture can be any size and shape; their purpose is to fill in a large volume of space. For instance, when I visited Egypt for the first time in 1986, a helpful man whom I ran across near the so-called Queen's Pyramids located south of the third pyramid (Kephren's) showed me an opening in the south side of the pyramid. At his insistence we crawled into it, and I found myself in a large space looking up at a massive core masonry block. Not knowing the full size of the block, what I could see on this block's underside exceeded by far the dimensions of the core masonry blocks that can be seen outside the Great Pyramid.

From Egypt to Lebanon, Turkey, and many other places around the world, our ancient ancestors demonstrated that it was routine to cut, transport, and erect massive blocks of stone, some weighing more than 1,000 tons. Aside from the structural strength obtained by using such large-sized blocks of stone, from an economic and efficiency perspective, the greater the volume contained in a single block to fill in a space the better, for the cost of cutting into smaller blocks is avoided.

Production Considerations

The volume of stone that makes up the core masonry greatly exceeds the volume that comprises the casing stones and interior architecture. If the percentage of precision-cut stone is, say, 20% of the total amount of stone, then core blocks that are five times the volume of precision-cut stone must be quarried, delivered, cut on site, and positioned so that, ideally, all layers of the pyramid rise at the same time. However, a reasonable amount of time between the first stone being laid and the last one may be tolerated. Also, the personnel on site can decide what size block is added to the core masonry. For this work, the parameters they are required to work within have already been established by the existing precision-cut stone that was tightly controlled by architects and metrologists.

Building the Tesla Dome

Positive electron holes that are squeezed out of the igneous rock and flow upward to and through the electron harvester will combine with negative electrons in the Tesla dome, where it will radiate out in the same manner as Tesla imagined, except unlike Wardenclyffe Tower, which received power from a conventional power plant and distributed it wirelessly, the pyramid electron harvester dome is receiving electrons directly beneath its location on the earth's surface. As Friedemann Freund's research showed, this process is initiated when the igneous rock is subject to pressure. Even after the pressure is released, the positive holes continue to flow. To measure this, Freund created a circuit between the end of the granite being put under pressure and connected it with a wire to a copper cap at the other end.

It may be argued that no evidence exists to indicate a dome once covered the Great Pyramid, or any other pyramid for that matter. It's true. There is nothing to see in the archaeological record or in any hieroglyph or relief that I'm aware of that shows a pyramid covered by a dome. Nonetheless, if we were to build a dome over the Great Pyramid in Egypt, Petrie identified four sockets near each corner that we might use as anchor points:

> Since the time of the first discovery of some of the sockets in 1801, it has always been supposed that they define the original extent of the pyramid, and various observers have measured from corner to corner up them, and thereby obtained a dimension which was— without further inquiry—put down as the length of the base of the pyramid. But, in as much as the sockets are on different levels, it was assumed that the faces of the stone placed in them rose up vertically from the edge of the bottom, until they reached the pavement (whatever level that may be) from which the sloping face started upwards. Hence it was concluded that the distance of the socket corners was equal to the length of the pyramid sides upon the pavement.
>
> On obtaining accurate measures, however, of the relations of the sockets to the casing on each side, it was found that the sockets lay two or three feet outside the line of the casing of the pyramid on the pavement; and also that the mean planes of the core masonry of the Pyramid were far more nearly a true square than the square of the sockets. The socket distances varying on an average 4.4 inches from the mean and the core size varying but 1.0 inch from their mean length; while there was also a similar superiority in the squareness of the core. This first threw doubt on the sockets representing the original base; and on comparing that distance from the center of the pyramid. It was seen that the deeper the level of the socket the further out it is from the center (Petrie 1883, 10).

If it is decided that a dome would add value and functionality to power distribution, it can be designed to provide multiple connection points on each face of the pyramid and also the top. While the material the dome is made of and its design are yet to be determined, when considering how it is constructed and how quickly, there are several build plans that could be considered. See color plate 6.

Ideally, the dome will grow at the same time the pyramid is being built, and the dome will reach the top not long after the pyramid capstone is put in place and will be connected to it. Using technology that is available at this time, there are several ways of achieving this.

- Use traditional fabrication out of foundry/mill metal. Sections are shipped to the site and assembled.
- Use 3-D printer robots to create the dome on site as a monolithic structure that does not have joints.
- Print interlocking modules that will be brought to the site and assembled. (Dare we even imagine that nanomachine universal assemblers may have a hand in this?)

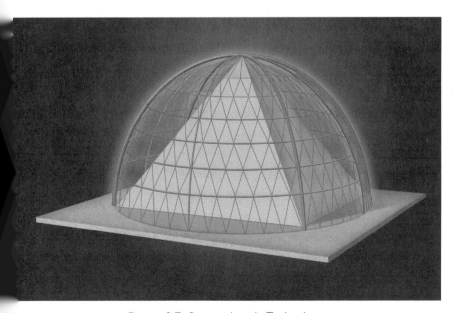

Figure 9.7. Pyramid with Tesla dome

I wouldn't even dare to imagine the advances in construction technology and capabilities we will have in 20, 30, or 50 years. Nor did I imagine 25 years ago the advances in construction that today incorporate functional capacity with elegant and beautiful design, such as the Museum of the Future in Dubai. I would anticipate that when it is time to build the pyramid electron harvester, considering the advances made in 3-D printing in the last 10 years, its application will reflect all knowledge, imagination, and skill in a dazzling display of beauty that will shock and inspire the world.

Operating Costs

Considering that the electron harvester would be located directly above the energy source, its operating costs would be miniscule compared to the operating costs for a conventional power plant. After it is built, no more material, other than chemicals, needs to be extracted from the Earth. It won't need to burn coal, oil, or gas or use nuclear material to generate electricity. However, it's impossible to calculate accurately what the energy output and associated operating costs will be at this time. Amortized over a fraction of the life of the harvester, I predict that the ROI will be relatively short. Six million tons of precisely cut rock delivered, assembled, and functioning in such a way as to squeeze electrons out of the Earth would run for years, maybe decades, with little interruption. Materials for continued operation, such as chemicals for the Queen's Chamber shafts and perhaps oil for maintaining the pulse-generation equipment in the Subterranean Chamber would need to be sourced and delivered. Then there would be site maintenance and administrative costs, but that's about it.

Environmental Impact

The environmental impact of the pyramid electron harvesters will be transformative, not just in how the biosphere will be improved but also in the effects that it will have on how humans and animals respond to an environment that exists in harmony with the Earth and the universe. Acoustically and visually, our environment has changed since the dawn of the industrial revolution. We have learned to live with discordant noise and bad smells. Developing machines and systems that are in harmony with the universe's natural frequencies can benefit all of earth's living creatures.

Maintenance Costs

The electron harvester will require occasional maintenance. This is indicated by the evidence discovered at the upper end of the Queen's Chamber shafts. It is clear that the South Shaft electrodes would need to be changed occasionally; the question is, how is that area accessed to change them? I had previously discussed the possible existence of access tunnels being included in the design of the pyramid, similar to the winding tunnel proposed by Houdin, and that may be the case. However, the question Hany Helal asked me at the close of our meeting in September 2021 was one of those important questions that, like Tompkins's question that I discussed at the beginning of the book, opened my mind to what I had previously considered inconceivable and was a question from which this chapter grew in scope and detail.

RESTORING THE GREAT PYRAMID IN EGYPT

When my meeting with Hany Helal and Hamada Anwar was drawing to a close, Helal asked me if the Great Pyramid could be restored to operational status. I had thought about this question before but not very deeply, as I never thought that a senior member of the Egyptian government would even ask that question. Moreover, it was my personal opinion before this meeting that the Egyptian people would not support such an endeavor. Another factor was that calculating the costs and time to complete a full restoration is beyond my ability. It's complicated and I've had a lot of time to ponder that question since it was asked, but after Helal asked it and I had become aware that attitudes were changing not only among the population in Egypt but in the highest level of Egyptian government, the question deserved a more serious study.

That this section is the last one in this chapter is not an indication of its level of importance. In fact, it is the singular question that inspired this chapter to be written. It is only last because of the sequence of engineering steps—design, build, operate, and maintain. In many ways, it is probably the most important section of the chapter.

As I was writing this chapter, in the back of my mind was the thought that some readers are probably asking, "How are the blocks that are being placed in the pyramid electron harvester going to be transported and lifted into place?" I believe I know the answer to that question. Just like the questions asked as more information has been provided on the Great Pyramid over many years, it relies on reverse engineering, but it is also different because, with the Great Pyramid we have been able to study it closely inside and out and have amassed an enormous amount of data, but we have not yet seen it operate.

With the recent disclosure of UAPs and the acceptance of the fact that machines are navigating within our airspace and exhibiting extraordinary gravity-defeating capability, we are witnessing a machine's operation but we have not been able to gather any data that explains the technology employed to accomplish it. Understanding the technology that allows these machines to perform as they do and using that knowledge to develop systems and techniques for eliminating the effects of gravity when lifting and moving massive blocks of stone is imperative both to the building of another Great Pyramid and repairing one—including the original in Egypt. No, I'm not suggesting that we cause a UAP to crash (even if we were able to) in order to have access to it. However, I think understanding how Edward Leedskalnin built Coral Castle would be time and money well spent. The most important lesson in all this is gaining an awareness that antigravity has been achieved in the past, is being demonstrated today, and can be achieved in the future.

To bring the Great Pyramid in Egypt back to operational status would require us to access its broken and damaged features and repair or replace them with the same care and precision as when it was originally built. For maintenance, repair, and restoration purposes, the ancient Egyptians would have been as capable and efficient in using their technology to take the Great Pyramid apart as they were when they put it together. In the event that the electrodes needed to be replaced at the end of the Queen's Chamber South Shaft, a long, winding tunnel within the mass of the Great Pyramid would not be feasible if just the removal of a few blocks on the south face would give us access to a short tunnel. With respect to antigravity, to my knowledge, we are not quite there yet. I don't expect we will get there in my lifetime, but I believe present and/or future generations will.

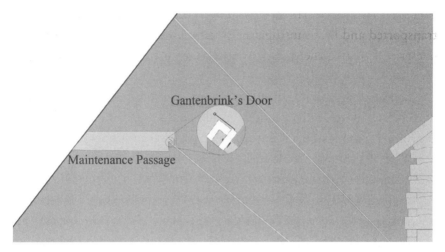

Figure 9.8. Maintenance passage to replace electrodes.

While the reasons for the angles of the Queen's Chamber South and North shafts have generally been interpreted to be symbolic—with their relationship to particular stars being an appealing and popular idea—I would like to propose a more practical reason for these angles. Two requirements were achieved when these shafts were oriented in this way.

1. Maintain a head pressure to provide chemicals to the Queen's Chamber in a specific amount over a period of time.
2. Position the end of the shaft at a point where it is close to the outside of the pyramid and more easily accessible for maintenance.

By having a vertical shaft, we can achieve one requirement but not both. By having a horizontal shaft of the same length, we cannot achieve either requirement.

In *The Giza Power Plant*, I addressed the question of how great civilizations come to a violent end. In 1998, both the US and Russia possessed significant nuclear arsenals, but there was relative peace, which made us more comfortable in the continuation of this status quo.

Considering the current unrest in the world, it seems appropriate to end this book the same way I ended my first book, though when I started writing this book in 2021, I believed it was an external force, such as asteroids, meteorites, or coronal mass ejections that caused cataclysmic events that wiped out prehistoric civilization, and wasn't thinking about nuclear:

One of the most inconceivable events with which modern humans are faced is nuclear disaster. Though the threat of all-out nuclear war between the United States and the former Soviet Union has been greatly reduced, it is still possible that our civilization could be wiped from the face of the Earth by a few miscalculations in foreign policy, a reckless terrorist act, or an error or malfunction in our own nuclear weapons or devices—the ones supposedly protecting us from a premature reaction to a nonexistent threat. Could it happen? Most of us believe that we, as a species, are simply too smart for these possibilities to overtake us.

Has it happened before? Were the ancient Egyptians smart enough to ensure that their own civilization would endure? The greatest lessons regarding our own mortality may begin with the pyramids of Egypt, the strong evidence of advanced machining practiced by the ancient Egyptians, the geological and biological records, and the world's ancient sacred records. These are all pieces of a giant puzzle that so many of us are trying to piece together. I have hope that we will regain this lost knowledge and learn from the lessons of the far distant past in time to save our own society from the fate that likely befell advanced civilizations that came before us. And I hope that along with granting us the wisdom to survive, this knowledge also may provide us the means through which we can evolve–spiritually, intellectually, and technologically—into more than we have ever chanced to dream (Dunn 1998, 256).

A BRIEF EXPLORATION INTO THE ACOUSTICS OF THE GREAT PYRAMID

ROBERT VAWTER

INTRODUCTION

Much has been written on the sound (acoustics) within the Great Pyramid. Unfortunately, there have been no thorough systematic measurements of the pyramid to substantiate the claims made by most of those authors. This article summarizes a few of the better researched attempts to examine the acoustics of the pyramid and presents my own research over the last 25 years.

Other UNESCO sites, such as the Hagia Sophia and the Taj Mahal, have been examined by acoustics experts. In the case of the Hagia Sophia, Stanford University's Center for Computer Research in Music and Acoustics (CCRMA) developed a computer simulation and processing program to emulate the acoustics. While the measurements conducted by Thomas Danley are deservedly cited often, his data is limited to the King's Chamber. Unfortunately, Danley published precious little specific frequency information (Danley 2000). No systematic study of the entire pyramid's acoustic environment using modern techniques, such as Fast Fourier Transform (FFT) analysis of sine wave sweeps, has been done as of the time of this writing.

NOTES FROM THE RED PYRAMID AT DAHSHUR

Back in 1996, I was part of an experiment in the Red Pyramid at Dahshur, along with Abd'el Hakim Awyan and Stephen Mehler. It was a very rough technique using chanting as the sound source and measuring the resonant frequencies when we found them. I used a Korg professional tuner certified to be accurate to 1 cent, which is ¹⁄₁₀₀th of the difference between two notes in the modern equally tempered scale.

While in the pyramid, we measured 26 resonant frequencies. Later that evening, I noticed only six different frequencies had actually been measured. Five "notes," or octaves of them, had been measured several times at different places, and one frequency had measured only the upper chamber. At the time, there was a large excavation in the floor that has since been covered up. I subsequently decided to discard that single outlier frequency as not being part of the original design specifications. The dominant frequency seemed to be an F# plus 18 to 23 cents, as we had several solid readings in that very narrow range. This "note" is slightly less than a quarter of the way between a modern F# and the modern G. When I returned home, I connected with a friend who had a small harp, and we tuned the harp to the five main frequencies measured in the pyramid, omitting the outlier from the room with the large excavation in the floor.

To our surprise, the five notes were harmonically related and formed a musical scale. It was not a scale in the modern equally tempered sense—in other words, the frequencies did not align with the modern notes A, B, C, etc. Yet when played it was very pleasant sounding. This observation is analogous to examples of other ways to construct musical scales today, such as the variety of scales found in Japan, India, the Middle East, Native American flutes, and modern microtonal compositions. This experiment led to three fundamental realizations.

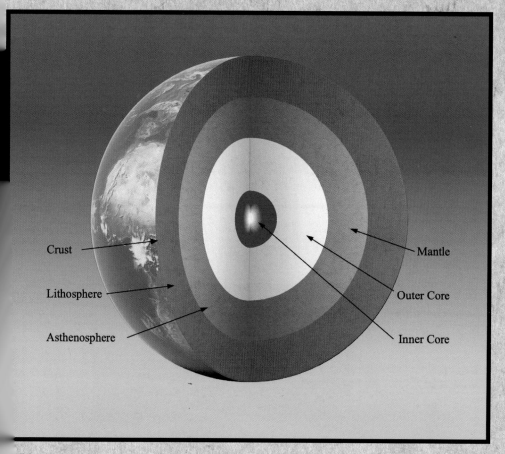

Plate I. *The layers of planet Earth. Earthquakes occur when the tectonic plates that make up the Earth's crust shift with the motion of the mantle. NASA physicist ʃiedemann Freund, PhD, developed a theory to explain the atmospheric phenomenon f earthquake lights—an electrical anomaly occurring in the Earth before and during rthquakes. Certain types of igneous rocks found in Earth's mantle and crust contain 'ormant electronic charge carriers. Freund's fundamental discovery found that when stressed, such as under earthquake conditions, these rocks hold an electric charge and essentially turn into a battery. Freund's theory was tested in laboratory conditions by himself and others and proven to be correct.*

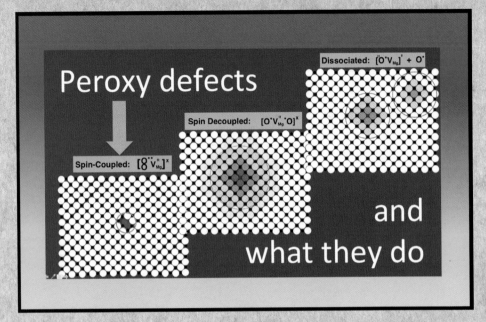

Plate 2. *Freund's illustration of the physics behind earthquake lights. Image credit: F. Freund (see Appendix B).*

Plate 3. *Freund's Laboratory at NASA where he successfully tested his theory. Image credit: F. Freund (see Appendix B).*

Stress build-up

Plate 4. *Freund's granite shown vertically to illustrate the same orientation at which igneous rock in the Earth is squeezed prior to and during an earthquake. Image credit: F. Freund (see Appendix B).*

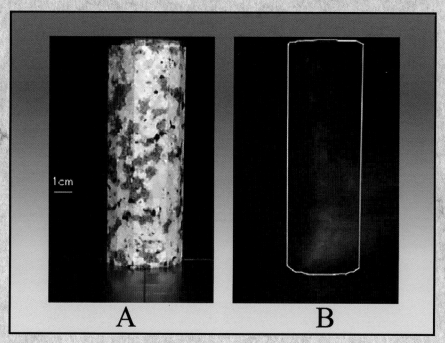

1cm

A   B

Plate 5. *Photographs of granite under stress.
A: Lights on; normal digital camera. B: Lights turned off; infrared camera.
It has been speculated that airflow in the room, flowing left to right,
caused the ions cloud to extend outside the granite.
Image credit: F. Freund after M. Kato (see Appendix B).*

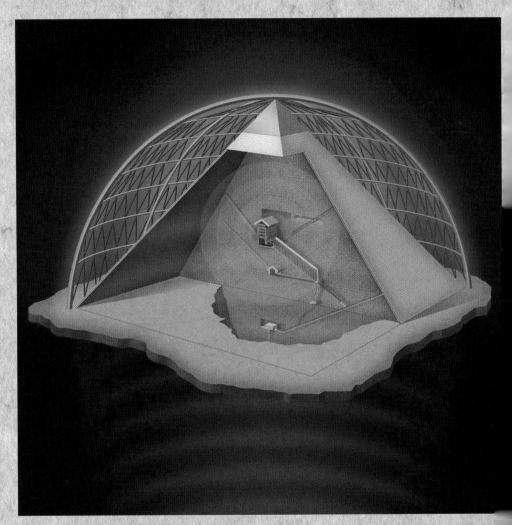

Plate 6. *Fulfilling the need to create a circuit, perhaps a dome could serve to radia[te]*
power into the atmosphere, as proposed by Nikola Tesla in his vision for the wireles[s]
distribution of electricity (Wardenclyffe Tower Project).

Plate 7. *Electron harvesters at Giza, Egypt. Electrons that are stimulated to flow out of the lithosphere will seek the highest point on the local topography. With the Giza Plateau serving as a pulse generator to cause vibrations in the lithosphere, the pyramids attract and focus the electrons that are squeezed out of igneous rock below.*

Plate 8. *The Pyramids of Giza: Manmade mountains of stone designed to attract electrons that are squeezed out of igneous rock within the Earth's crust.*

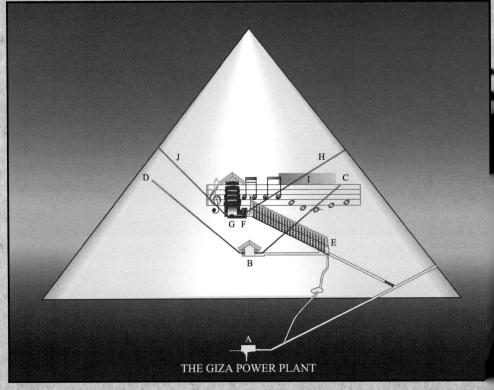

Plate 9. *The schematics of a solid-state power plant.*
A: Pulse generator [aka Queen's Chamber]
B: Hydrogen production chamber [aka Subterranean Chamber]
C: Chemical inlet conduit (hydrated zinc) [aka Queen's Chamber North Shaft]
D: Chemical inlet conduit (dilute hydrochloric acid) [aka Queen's Chamber South Shaf
E: Acoustic amplifier [aka Grand Gallery]
F: Acoustic filter? [aka Antechamber]
G: Resonant Cavity [aka King's Chamber]
H: Waveguide [aka King's Chamber North Shaft]
I: Microwave preamplifier [aka The Big Void]
J: Power output shaft [aka King's Chamber South Shaft]

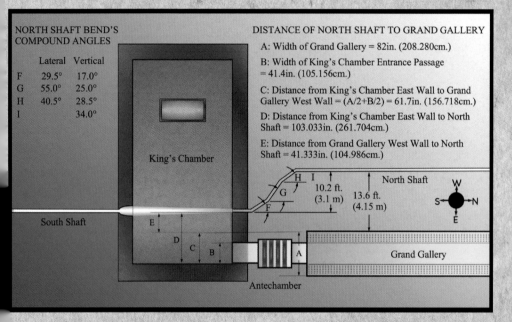

NORTH SHAFT BEND'S COMPOUND ANGLES

	Lateral	Vertical
F	29.5°	17.0°
G	55.0°	25.0°
H	40.5°	28.5°
I		34.0°

DISTANCE OF NORTH SHAFT TO GRAND GALLERY

A: Width of Grand Gallery = 82in. (208.280cm.)

B: Width of King's Chamber Entrance Passage = 41.4in. (105.156cm.)

C: Distance from King's Chamber East Wall to Grand Gallery West Wall = (A/2+B/2) = 61.7in. (156.718cm.)

D: Distance from King's Chamber East Wall to North Shaft = 103.033in. (261.704cm.)

E: Distance from Grand Gallery West Wall to North Shaft = 41.333in. (104.986cm.)

King's Chamber

South Shaft

North Shaft

10.2 ft. (3.1 m)   13.6 ft. (4.15 m)

Antechamber

Grand Gallery

Plate 10. *Top view of power center. Drawing by C. Dunn after R. Gantenbrink with dimensions from Gantenbrink (angles) and Petrie (lengths).*

The significance of the North Shaft, with its multiple compound angle bends cannot be overstated. Before it was fully explored, these inexplicable features in the Great Pyramid were generally thought to be air-shafts. Learning more about them has prompted some to awkwardly suggest they were/are "star shafts" or more esoterically "soul shafts" through which the King's soul was directed towards the heavens. Unfortunately, none of these theories can explain the complications designed into the shafts. Nor do these theories explain why the shafts could not merely run straight to the outer surface. Is it possible that the Grand Gallery wall blocks are almost 13 feet (4m) thick? There is no other way to interpret this evidence other than it exists to serve a much higher purpose.

The component parts are:

- Grand Gallery (Acoustic Amplifier)
- Antechamber (Acoustic Filter or a Thermo Acoustic engine stack, as proposed by Eric Wilson [see Appendix C])
- Kings Chamber (Resonant Cavity)
- North Shaft (Microwave signal input)
- South Shaft (Power output)

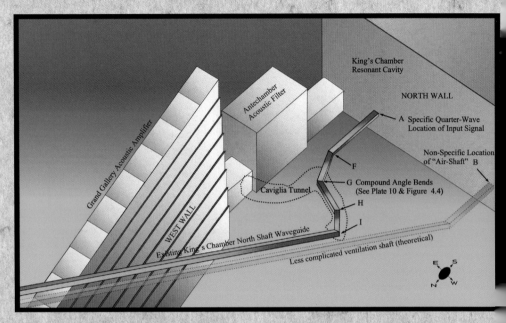

Labels within image:
- King's Chamber Resonant Cavity
- NORTH WALL
- Antechamber Acoustic Filter
- A Specific Quarter-Wave Location of Input Signal
- Non-Specific Location of "Air-Shaft" B
- F
- G Compound Angle Bends (See Plate 10 & Figure 4.4)
- H
- I
- Grand Gallery Acoustic Amplifier
- WEST WALL
- Caviglia Tunnel
- Existing King's Chamber North Shaft Waveguide
- Less complicated ventilation shaft (theoretical)
- E S N W (compass)

Plate 11. *Mechanical complications that are designed into a product are engineered according to required outcomes. For instance, the more features a mechanical timepiece has, the more complicated its design and manufacture becomes.*

While the King's Chamber and Queen's Chamber shafts have been described as "air shafts" "star shafts" or "soul-shafts" for the King to use to exit the pyramid tomb and travel beyond, the evidence argues strongly against these theories.

Shaft A could be directed to the outside passing the Grand Gallery at a distance of 41.3 inches (104. 986cm) with just one bend. But it has four. Complicated. Why? Shaft B is a theoretical air shaft that only has one bend.

Solution: Fingerprints of an ancient waveguide.

Waveguide enters the Resonant Cavity at the quarter-wave point where a standing wave in a resonant cavity has the highest amplitude. Similar to modern waveguide design, multiple bends are used. The Waveguide passes the Antechamber Acoustic Filter and Grand Gallery. See Chapter Four and Plate 10. Drawing by C. Dunn base on R. Gantenbrink's wire-frame CAD drawing.

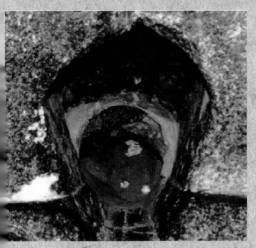

Plate 12. *Opening in south wall of Resonant Cavity leading to South Shaft. I first observed the unique microwave horn antenna-like features of this opening and photographed it in 1986 (left image). Returning in 1995 and hoping to be able to gather geometric data using a die sinking profile gage, a ventilation fan, installed by Rudolph Gantenbrink (circa 1993), presented a barrier (right image).*

Plate 13. *North Shaft opening in the north wall of the Resonant Cavity. Dimensions of 8.4" wide by 4.8" high make it suitable for a hydrogen microwave signal with a wavelength of 8.309 inches (21 cm). Note: North and South openings are directly opposite each other.*

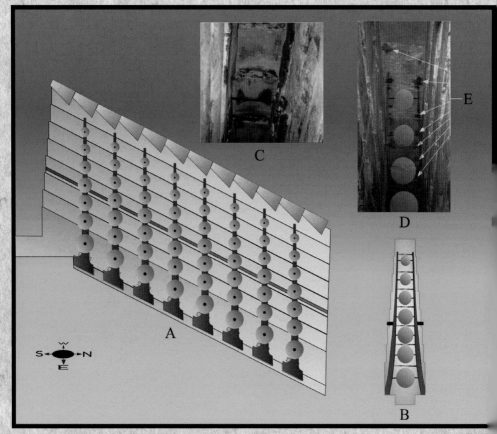

Plate 14. *Section of acoustic amplifier showing 8 of 27 resonator stacks.*
A: East-West Cross Section. Resonator frames are locked into position using recesses cut into the side ramps. Vertical orientation is established by using the slots cut into the corbel along the length of the east and west wall.
B: North-South Cross Section.
C: Close up photograph of scorch marks on ceiling of Grand Gallery after the interior of the Great Pyramid was cleaned. Photograph taken in 2001.
D: The Grand Gallery illustrating the correlation between the pairs of scorch marks (E) and the resonator frames that were first discussed in
The Giza Power Plant, 13 years earlier.

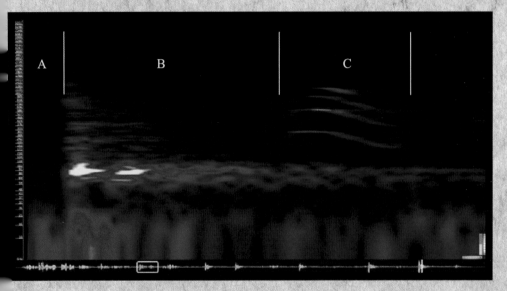

Plate 15. The powerful acoustic properties of Great Pyramid's inner chambers and passages have long been of intense interest. In 2019, sound engineer Robert Vawter, using a Tascam recorder, performed a simple test by stomping on a loose board near the lower end of the Grand Gallery. This image shows the resulting acoustic spectra. (See Appendix A for more details).

A: Gallery is silent except for low pulsing infrasonic frequencies at around 5–20 hertz. These frequencies continue through the duration of the recording.

B: Vawter stomps on the loose board, resulting in a broad-spectrum response from the Gallery's corbels.

C: Vawter's wife Helen says "WOW" when hearing the reverberation.

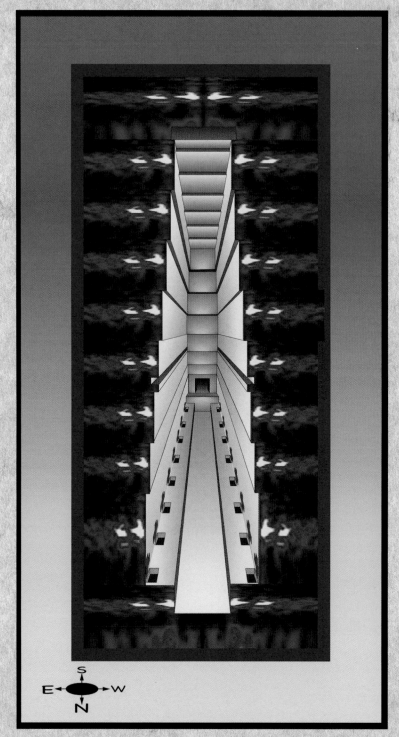

Plate 16. *Robert Vawter's spectra with Grand Gallery illustrating corbels responding at different frequencies to sound input. Note: This graphic is for visual effect and is not intended to suggest that the acoust spectra gathered inside the gallery exists within the limestone wall blocks.*

AN AMAZING MASER

Plate 17. *The King's Chamber shaft positions suggest a flow of energy and maser (microwave amplification through stimulated emission radiation) activity. Electrons are stimulated to flow not just from the igneous rock within the Earth but also from the thousands of tons of igneous rock (rose-red Aswan granite) from which the power center (aka King's Chamber) is constructed.*
Vibrating mass of granite + "Freund Effect" + hydrogen = pollution-free power plant.
A: Microwave signal input.
B: Resonant Cavity containing highly energized hydrogen.
C: Similar to how a laser operates, except in the microwave region of the spectrum, the input signal stimulates emission from the energy-pumped hydrogen, which then exits the power center through a south shaft that has an opening similar to a microwave horn antenna.

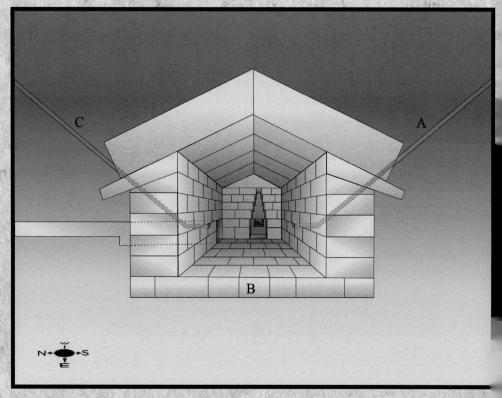

Plate 18. *Hydrogen production chamber (see Appendix D).*
A: Chemical inlet shaft.
B: Mixing vessel.
C: Chemical inlet shaft.
Note: The author proposed that the chemicals used were hydrated zinc and dilute hydrochloric acid were mixed to produce hydrogen, which flowed into the Resonant Cavity and pumped to a higher energy level by acoustic and electromagnetic energy. Appendix D, written by Dr. Brett Cohen, present other alternative solutions for creation hydrogen.

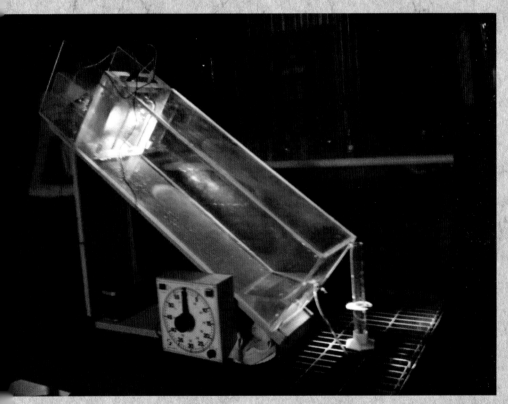

Plate 19. *Author's model of chemical delivery shaft to demonstrate function of metal pins in the blocking stone at the end of the shaft. The experiment was successfully demonstrated to work and was recorded by Ancient Aliens for the History Channel. The shaft was filled with water until it covered the pins, at which time it created a circuit which turned on a light. This photograph was taken after the tank had drained and the light was turned on for visual effect.*

Plate 20. *Left to right. Robert Vawter, Chris Dunn, and Friedemann Freund at the SETI Institute. (See Appendix A and B.)*

Plate 21. *Eric Wilson in Egypt (see Appendix C). Image credit: Lauren Kurth Eric Wilson.*

1. There is no good reason to assume the ancients used modern musical sensibilities. This view is supported by looking at research into the Ney flutes by Professor Mahmoud Effat, Professor Fathi Saleh, and Professor Robert Gribbs. Their research found microtonal relationships as well as a diatonic scale in the replicas of the ancient flutes they tested. Additionally, changes in the tunings of the flutes illustrated the evolution of music during dynastic times (Effat 2000).

2. Resonant frequencies typically are emphasized by the presence of parallel surfaces and the formation of standing (resonant) waves bouncing back and forth between those surfaces. In acoustics, these are known as modes or eigenvectors. The frequency in these cases is determined by the distance between the walls. In short, the dimensions of the rooms and the distance between the corbels in the Red Pyramid determined the resonant frequencies we measured. Since those frequencies did not correlate to the modern equally tempered scale, it was necessary to deal with the acoustics in terms of frequency only, and we were not able to draw parallels to modern Western music.

3. The realization that we were looking at a musical scale led to a rather unexpected conclusion. To construct a building the size and complexity of the Red Pyramid and to have the resonant characteristics precise enough to create melodic relationships between the individual notes could not have been accidental; it must have been intentional. In other words, the ancient Egyptians knew what they were doing.

THE GREAT PYRAMID

Any discussion of resonant frequencies must include the dimensions of the rooms, corridors, corbels, etc. These are the primary determining factors for resonance. In the case of the pyramid, the hard rock surfaces emphasize this point. While modern laser scans have been done, the data is unfortunately not publicly available. The estimates cited in this paper primarily use the measurements provided by Sir William Flinders

Petrie: "Probably the base of the (King's) chamber was the part most carefully adjusted and set out; and hence the original value of the cubit used can be most accurately recovered from that part. The four sides there yield a mean value of 20.632 ± .004, and this is certainly the best determination of the cubit* that we can hope for from the Great Pyramid" (Petrie 1883, 28).

This measurement, and multiples of it, repeat throughout the Great Pyramid, and will be referred to often in this paper. For example, the Petrie cubit x 10 equals the width of the King's Chamber and is exactly half of the length of that chamber. This measurement is also found in the Queen's Chamber and in the spacing of the Girdle Stones found in the Ascending Corridor.

The repetitive nature of certain frequencies throughout the pyramid leads to another observation.

In my experience setting up and operating large sound systems for concerts and theatre, it was necessary to identify resonant characteristics in rooms and take steps to control or remove them before problems such as feedback occurred. This could ruin a performance. There are also related issues regarding reverberation, intelligibility, and psychoacoustics that affect a complete description of an acoustic environment. However, the pyramid seems to be designed to do the exact opposite— deliberately repeating and emphasizing certain frequencies throughout the entire structure. In other words, it is creating a very specific acoustic environment, which amplifies certain frequencies in multiple aspects of the design. The interaction of these waveforms is known as nonlinear acoustics.

Only certain parts of the pyramid's internal structure will be discussed here, as the existence of the blocks in the lower part of the Ascending Corridor would have isolated the Descending Corridor down to the Subterranean Chamber from the rest of the system. Similarly, the Queen's Chamber Corridor and the Queen's Chamber itself will only be mentioned briefly, as there is evidence that they were also isolated from the Grand Gallery (Maragioglio and Rinaldi 1965).

---

* One Royal Egyptian cubit = 20.610 inches (0.523 m)

These limitations also focus the discussion on the system Chris Dunn is describing in his book (Dunn 1998).

THE ASCENDING CORRIDOR

To determine approximate resonance characteristics of the Ascending Corridor in the pyramid, I have modeled it as a long, rectangular organ pipe. These equations are well established, and additional phenomena, such as end correction, can be included. It is also possible that bell tones may be generated, since there is a slight tapering in the corridor as one moves down toward the granite blocks, which could also be interpreted as a slight flare as the corridor opens up into the Grand Gallery. Additionally, the existence of the blocks in the lower section determined the corridor should be treated as a closed organ pipe. This significantly changes the harmonic content of the resulting sound.

It is difficult to estimate the exact length from the end of the blocks to the top of Ascending Corridor. Petrie concluded that the end of the present top block is not in the original position: "The granite plugs are kept back from slipping down by the narrowing of the lower end of the passage. . . . The present top one is not the original end; it is roughly broken, and there is a bit of granite still cemented to the floor some way farther South of it. . . . From appearances there I estimated that originally the plug was 24 inches beyond its present end" (Petrie 1883, 21).

Taking the overall length, and allowing for Petrie's observation, gives an approximate length of 107.8 feet (32.86 meters) from the end of the blocks as originally placed to where the corridor opens into the Grand Gallery. These dimensions yield an estimate of 2.6 Hz for the fundamental frequency and 7.8 Hz for the third harmonic for an organ pipe of this size. This is well below the threshold for human hearing. Chris Dunn has theorized that the pyramid is acoustically coupled to Earth's natural vibrations (Dunn 1998, 138, 219). While the estimates for the Ascending Corridor frequencies are higher than the extremely low frequencies (ELF) discussed in the ocean-caused scientific literature on earth hum, this frequency range is consistent with seismic activity.

A paper published by the United States Geological Survey on earthquakes and geologic hazards states the following: "When a fault ruptures, seismic waves are propagated in all directions, causing the ground to vibrate at frequencies ranging from about 0.1 to 30 Hertz. Buildings vibrate as a consequence of the ground shaking. . . . Compressional waves and shear waves mainly cause high-frequency (greater than 1 Hertz) vibrations which are more efficient than low-frequency waves in causing low buildings to vibrate" (Hays 1981).

The Descending Corridor is long enough to get into the ELF ranges. At about 345 feet (105.156 m) long, an organ pipe of that size could be expected to produce a fundamental frequency of about 0.82 Hz. Since infrasound travels though buildings and solid rock, the pyramid could be reacting to these frequencies. Only specialized equipment such as geophones would let one know if that is the case here.

I believe this supports Chris Dunn's theory that the pyramid may have been acoustically coupled to Earth frequencies by employing the physical properties of tuned resonance. This line of thinking deserves continued investigation.

THE GIRDLE STONES IN THE ASCENDING CORRIDOR

One problematic detail in the pyramid design deserves mention here. Standard pyramid theory states the so-called Relieving Chambers (A) function as a stress-relieving feature to mitigate the weight from the blocks above the chamber and prevent the King's Chamber roof from collapsing. If this is the case, then why is there no analogous feature over the Queen's Chamber (B)? It is lower in the pyramid and closer to the centerline, so there is significantly more weight pressing on it. The angled roof is sufficient there, so why was it deemed necessary over the King's Chamber?

A similar situation can be seen in the Ascending Corridor. The presence of the three Girdle Stones near the center of the corridor (D) is also classified as a stress-relieving feature designed to prevent the passageway from collapsing. However, the weight above the corridor is

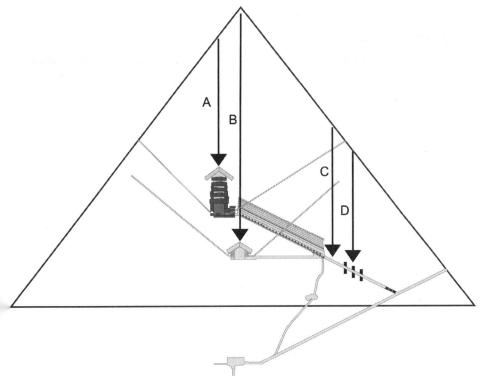

Figure Aa.1. Weight distribution locations discussed in this section.
Appendix A images credit: Robert Vawter unless otherwise noted.

greater in the section (C) farther up the passage from the location of the Girdle Stones. This is due to the outer slope of the Pyramid rising at a rate approximately double the slope of the Ascending Corridor floor. If the Girdle Stones were not needed for structural reasons, could there be another explanation?

The Girdle Stones are equally spaced "on center" like studs in a wall. This is the same measurement found in the King's Chamber width (206.32 inches [5.24 m]), and equal to half of the King's Chamber length. This measurement also appears in the Queen's Chamber and is directly related to other dimensions found in the Grand Gallery.

In short, my acoustic analysis is to view the "corridor" as an infra-sonic, closed-end organ pipe rather than classify its function as a corridor or passageway. To see this analogy in more detail, it is necessary

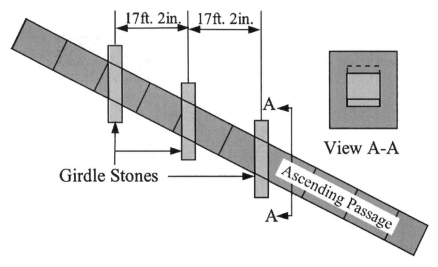

Figure Aa.2. Detail of the Girdle Stones in the Ascending Passage (Edwards 1975, 122; Kingsland 1996, 68–69). Graphic by C. Dunn.

to look at the precise measurements of several features in the Ascending Corridor. William Kingsland compared Petrie's measurements and those made by Morton Edgar (Kingsland 1996). Kingsland also draws attention to the deviations in the masonry layout in the corridor. His detailed drawings show some specific seams are a true vertical, while others are angled perpendicular to the slope of the floor.

Starting from the granite blocking plugs and going up the corridor, there is a small area before the first vertical seam in the walls. In the organ pipe analogy, this would correlate to an area known as the end correction zone, where the sound wave is not completely coherent due to interference from waves reflecting off the closed end of the pipe. This phenomenon makes the pitch sound slightly flat compared to strictly calculating the frequency (pitch) as a simple function of the length.

Moving up the corridor from there, five more vertical seams can be seen before the corridor ends at the opening to the Grand Gallery. Three of these seams are where the Girdle Stones are located. These features are equally spaced, within a variation margin of about a quarter inch. The remaining length of the corridor is precisely twice the on-center spacing as seen in the Girdle Stones.

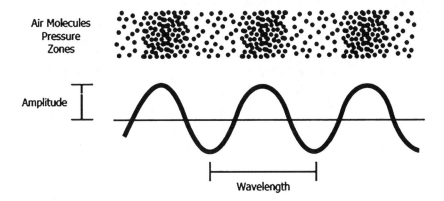

Figure Aa.3. Pressure zones within the Ascending Corridor.

If one visualizes a resonant-sound sine wave passing through this organ pipe (see Figure Aa.3), these vertical seams and the Girdle Stones are located within half an inch (1.27 cm) of the estimated location of the nodes where pressure waves would impact the side walls of the corridor.

In the organ pipe analogy, calculating the frequencies for multiple nodes within the pipe yields an interesting correlation. Six nodes generate a frequency around 33.9 Hz, a frequency very close to a King's Chamber axial mode length calculated at 32.7 Hz. Similarly, the calculation for twelve nodes in the pipe equals 65.2 Hz, which compares very favorably with the main resonant frequency, around 65.4 Hz, for the width of the King's Chamber. If one considers end correction in the estimates, which makes the pipe sound slightly flat (a lower frequency), these numbers become even closer to alignment. Is this a coincidence or intentional? I suggest the Girdle Stones' function is to provide a solid reflective surface to emphasize and/or stabilize this particular frequency.

THE GRAND GALLERY
AND THE RESONATORS THEORY

Once again, standing wave resonances are primarily found in parallel walls. In the case of the Grand Gallery, the only parallel surfaces are the

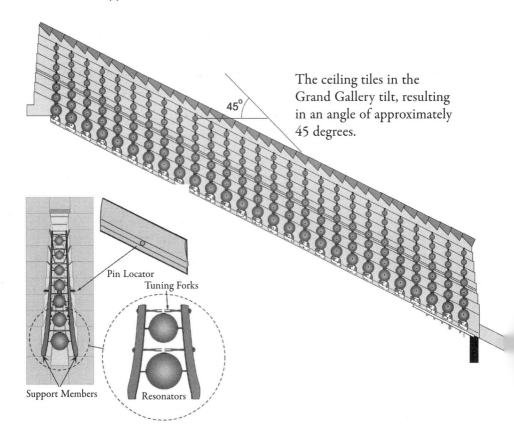

The ceiling tiles in the Grand Gallery tilt, resulting in an angle of approximately 45 degrees.

45°

Pin Locator
Tuning Forks
Support Members   Resonators

Figure Aa.4. The Grand Gallery and resonator arrays as suggested by Dunn.

corbels, specifically the side-to-side orientation in the gallery. Since the roof has a sawtooth construction detail, it serves to reflect sound down into the Gallery, where Dunn postulates resonators may have been located. Dunn further suggests the resonators were placed between the corbels. This raises the question: At what frequencies would resonances naturally occur at these specific locations (Dunn 1998, 168)?

Since the data from laser scans made by the Time Scanners team in 2014 (featuring structural engineer Steve Burrows, Caitlin Stevens, Dallas Campbell, and others) is unavailable to the public, only very rough estimates of the corbel dimensions can be made. Extrapolating an estimate from Petrie's measurements and the beautifully detailed drawings of Vito Maragioglio and Celeste Ambrogio Rinaldi yields the following approximations.

Maragioglio and Rinaldi state that the blocks of stone in the walls are corbelled inwards by 2.4–3.9 inches (6–10 cm) on each side. Since the inset is on both sides, this would mean approximately 4.72–7.87 inches (12–20 cm) per level as one counts upward (Maragioglio and Rinaldi 1965).

Another way to estimate these dimensions is to use Petrie's measurements of the floor, including the channel and the ramps on each side (Petrie 1883, 25). The width of the top corbel (the roof) is the same as the channel. Dividing this into an equal inset for each of the level changes yields an average estimate of 5.89 inches (14.96 cm). This dimension falls comfortably within the range suggested by Maragioglio and Rinaldi. For simplicity, I have used the 5.89 inches to generate the following chart.

	Width	Frequency	Modern Music Note
Corbel 7 / Roof	41.26	327.4	E4 minus 12 cents
Corbel 6	47.15	286.5	D4 minus 43 cents
Corbel 5	53.05	254.6	C4 minus 47 cents
Corbel 4	58.95	229.2	A#3 minus 29 cents
Corbel 3 w/ slot	64.88	208.2	G#3 plus 5 cents
Corbel 2	70.74	191.0	G3 minus 45 cents
Corbel 1	76.63	176.3	F3 plus 17 cents
Floor level > Wall to Wall	82.53	163.7	E3 minus 12 cents

Figure Aa.5. Musical frequencies of the Grand Gallery corbels.

If the resonances of these "vases" were tuned to the corbel dimensions, it can be seen that the frequencies precisely frame an octave range from approximately 164 to 328 Hz. This octave is divided into eight "notes," including the channel and shelf areas. It should be pointed out that the roof, channel, and shelf area are all direct multiples of the Petrie cubit. The top end of this octave, 328 Hz, may also correlate with Thomas Danley's "very rough estimate of 300 Hz" for resonating the blocks in the King's Chamber ceiling (Danley 2000).

THE RESONATORS—VITRUVIUS AND *ECHEA* (RESONATING URNS)

Conversations with Chris Dunn over the years have indicated that there is some debate over ancient knowledge and use of tuned vases as resonators in large architectural structures. He has asked me to take a moment to discuss the writings of the Roman architect and engineer Vitruvius. In his seminal work, *De architectura*, book 5, chapters 4 and 5, Vitruvius provides an extensive description of *echea* (resonating vases) used in Roman and Greek theaters to control the frequency content of the sound and reverberation in the space for speech intelligibility reasons.

Vitruvius also provides insight into the lineage of the transmission of information regarding the proper usage of these resonating urns. In the text he names Aristoxenus as a source (Vitruvius 1914, 139–46). The Greek philosopher Aristoxenus (c. 375–335 BCE) is primarily known for his musical treatise, *Elements of Harmony*. His teacher, probably Xenophilus, clearly introduced him to Pythagorean doctrine. Xenophilus was a Pythagorean philosopher and musician who lived in the first half of the fourth century BCE. Tracing the transmission of this knowledge directly back to the Pythagorean lineage is significant.

Pythagoras (c. 570–495 BCE) was in Egypt some 175 years before Alexander the Great, the rise of the Greek Ptolemaic Dynasty, and the early formation of the famous Library at Alexandria or in other words, before Greek influences came to dominate Egyptian thought.

Iamblichus (c. 250–325 CE), in his book *De Vita Pythagorica* (On the Pythagorean Life), and Porphyry (c. 233–305 CE), in his book *Vita Pythagorae* (Life of Pythagoras), both tell us that Pythagoras was in Egypt for 22 years. That ended when Egypt was conquered by the Persian king Cambyses in 525 BCE and Pythagoras was captured and taken to Babylon.

Since no texts written by Pythagoras are known to exist, we do not know exactly what he studied during his 22 years in Egypt. Therefore, it is possible, but not proven, that this knowledge regarding the use of resonating urns originated in Egypt.

Getting back to the Grand Gallery, a brief comment on the saw-tooth roof design is in order. The angles present in the roof reflect the sound back down into the area where the resonators would have been. Additionally, when sound waves in air (pressure waves) come in contact with a hard surface, such as the limestone or granite inside the pyramid, there is no phase change upon reflection. Further, if one views the collection of molecules as a fluid, either air or water, the concept of energy density enters the discussion. Consequently: "The energy of the molecules reflecting off the wall adds to that of the molecules approaching the wall in the volume very close to the wall, effectively doubling the energy density and hence the pressure associated with the sound wave" (HyperPhysics n.d.). This phenomenon applies to the sound reflecting off the corbels as well.

Putting this all together, the corbels are tuned to a specific octave where the frequencies are all directly related to other significant features found in various parts of the pyramid. Further, certain physical phenomena appear to be utilized to amplify the intensity of these resonant waves. This is especially true in the confined space between the corbels where Dunn has speculated resonating urns may have been located.

RECORDING OF AN IMPULSE RESPONSE TEST INSIDE THE GRAND GALLERY

In 2018, I was on tour with Chris Dunn and performed an impulse response test inside the Grand Gallery. I found a loose board at the landing just above the entrance to the Queen's Chamber corridor and recorded a heavy foot stomp on the board with a Tascam DR-40 digital recorder.

Plate 15 is a screen capture from the spectral analysis program of the frequency distribution and the intensity of the acoustic response within the gallery. The spectrogram covers about 2.7 seconds of the recording. The vertical axis denotes the frequency and the horizontal axis is time. The colors denote the relative intensity of the sound at that specific frequency. The sweeping arcs on the upper right are my wife, Helen, saying "wow" in response to the impressive sound made by the foot stomp as it echoed around inside the pyramid.

Starting at the left side of the scan, there is a brief dark area just before the foot stomp, which is the vertical line with the main flare at approximately 88 Hz (yellow/red is louder than the blue areas), An additional early reflection in the mid-60s Hz range can be seen immediately forming as reflections begin to build. The Grand Gallery corbels' resonant octave frequency range, estimated earlier in this paper and covering the octave from approximately 164–328 Hz, can also be seen as discrete frequencies forming a vertical stack that dissipates rather quickly. An additional two octaves of overtones are also evident in the spectrogram.

Below the main foot stomp horizontal band there is a rather peculiar absence of sound, and below that there is a series of pulsing infrasonic frequencies. These cover a rather wide bandwidth from 5–20 Hz. Notice the pulse is present in the spectrogram before the foot stomp, then spikes for a fraction of a second during the foot stomp impulse. These infrasonic frequencies then continue to pulse strongly long after the initial decay time for the initial sharp sound stimulus. Examination of other sections of the recording indicate these infrasonic pulses may be essentially continuous, as they were present even when the Gallery was quiet.

The presence of this infrasonic pulsing suggests there is more research that needs to be done to explain the cause of the pulses. For example, city noise would appear as a relatively constant level of activity, not pulses as seen here. The wide bandwidth of these pulses also remains confusing at this time. Danley also commented on the presence of significant infrasonic activity in the King's Chamber. It remains to be seen if this can be connected to Dunn's assertion that the pyramid is designed to be a sympathetic resonator with the natural earth hum or other seismic sources.

THE ANTECHAMBER COMPLEX

I have not modeled this section of the pyramid. The antechamber complex, as it exists today, is basically a hollow shell of the original. Common theories include missing elements, such as movable slabs of rock intended to seal the entrance to the King's Chamber. Some authors, including Chris Dunn, have postulated that it functioned as an acoustic filter, while others think some other type of technology may have been mounted in there that was removed long ago. It is outside the scope of this paper to speculate on these theories, so I will move on.

As one leaves the Grand Gallery, the dimensions of the entrance to the Antechamber Corridor are nearly the same as the exit into the King's Chamber. Since the central room is much larger than the corridors, the chamber complex as a whole probably exhibits airflow changes in both pressure and velocity, in accordance with the Venturi effect. For example, leaving the Antechamber complex and heading into the King's Chamber, the passageway narrows significantly relative to the size of the Antechamber itself. This detail should result in an increase in the speed of the airflow, along with a concurrent lowering of the pressure. These changes will shift the frequencies of any sound wave passing through. Frequency is dependent on the temperature, pressure, and humidity of the environment. Additionally, this also affects the speed of sound in frequency calculations. While I have generally ignored detailing these variables elsewhere in this paper, this provides a good time to acknowledge a limitation on the estimates provided.

THE KING'S CHAMBER

It is common to hear researchers speak of a particular dominant resonant frequency in the pyramid. My research indicates there are a vast number of frequencies present, including many that are clearly related frequencies. It is Chris Dunn's thesis, and I agree with him, that the pyramid functions as a complete system, with interactions that become dauntingly complex to contemplate, let alone claim to completely understand.

To illustrate the point, Figure Aa.6 is a computer simulation of the first 4.5 octaves of resonant frequencies in a room the size of the King's Chamber. At first glance, the chamber itself appears to be a simple box, yet the complexity of the resonant environment is, hopefully, evident.

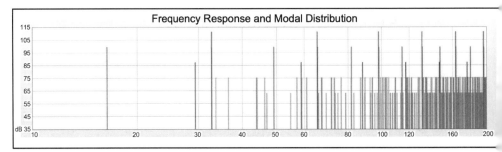

Figure Aa.6. Simulation of King's Chamber first 4.5 octaves.

In acoustic analysis, these resonant frequencies are known as room modes or eigenvectors. At the risk of boring the reader, perhaps a quick minute explaining what these are and how they work would be useful.

1. **Axial** modes tend to dominate the modal response. A rectangular room has three fundamental axial modes, which are formed between each pair of opposing wall surfaces.
2. **Tangential** modes involve any pair of parallel surfaces. They can bounce around all four walls, or two walls plus the ceiling and the floor. They are typically about half as strong as axial room modes.
3. **Oblique** room modes involve all six surfaces: all four walls, the ceiling and the floor. They are generally about one quarter as strong as axial modes and half as strong as tangential modes. Oblique modes often become overly excited in the low frequency (bass) ranges.

Room modes are also the main cause of sharp peaks and dips in the frequency response and can vary significantly depending where one stands in the room. The details of these interactions are known as nonlinear acoustics and can create significant amplitude changes in the acoustic environment. Further, a room with hard surfaces will exhibit "high-Q" characteristics, manifesting as sharply tuned resonances.

Axial Room Resonance Modes

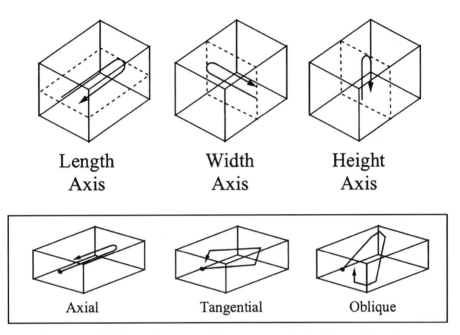

| Length Axis | Width Axis | Height Axis |

| Axial | Tangential | Oblique |

Figure Aa.7. The main modes (eigenvectors) in the King's Chamber.

These factors combine in ways that mean you might not hear certain frequencies at all, or hear them up to twice the volume, compared to a location a mere few feet away. These effects are most pronounced near the walls and in the corners.

This phenomenon is probably most applicable to the claims made by some authors that the coffer resonates in harmony with the King's Chamber. I maintain there is simply not enough data to know if this is true. It may be possible if there are significant additive waveform interactions in that part of the room. Conversely, the coffer resonance could be negated by specific interactions. The answer to these possibilities is dependent on one's location in the room, as was detailed in the above paragraph. Since the coffer has probably been moved from its original location, this line of inquiry may never be settled with any degree of certainty.

THE KING'S CHAMBER AMPLIFIERS
AND THE DANLEY 30 HZ EXPERIENCE

Using Petrie's measurements, it is clear that the chamber's length is twice the width. This results in a strong additive emphasis at certain frequencies. By definition, two sounds with either half or double the wavelength form notes an octave apart from each other. When these octave sound waves interact, the result almost doubles the amplitude of either wave individually. It seems possible the ancient architects were aware of this and chose to design the room to take advantage of this passive amplification effect. Researchers in a field known as archaeoacoustics constantly surprised by the sophistication found in ancient sites around the world.

Figure Aa.8 shows how this octave wave interaction would look on an oscilloscope. The top wave is the chamber's axial length frequency and the middle wave is the axial width frequency, exactly an octave higher. Notice the graphs show these waves to have an amplitude of +1 to −1. The bottom wave on the graphic shows the result of combining these waves. The combined waveform shows an overall loudness increase of almost +2 to −2, basically doubling the intensity of either wave by itself.

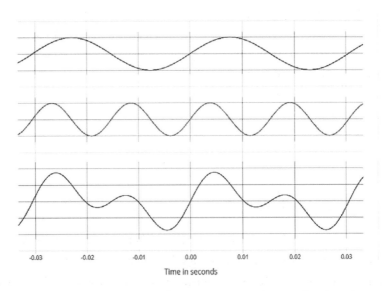

Figure Aa.8. Octaves. Graphic generated using Academo's
online beat frequency generator (Academo n.d.).

This may explain the comment made by Thomas Danley when a frequency "about 30 Hz" got so loud it caused the pyramid to vibrate and they had to shut down the experiment (Danley 2000). Various models indicate the axial width dimension is creating a resonance at about 32–33 HZ, which is emphasized by the octave of the axial length. Since axial modes carry the most acoustic energy in a room, I suspect they could form a significant resonant effect similar to Danley's experience.

Furthermore, this frequency and an octave of it are both present in the nodal analysis presented in the Ascending Corridor section and the Girdle Stone spacings. It seems plausible these two sections of the pyramid could interact and reinforce each other. If that was the case, it seems significant that Danley reported it shook the entire pyramid.

To conclude this section, I would like to draw attention to the relationship between the horizontal dimensions found on the floor (length & width) and the height of the chamber. The height of the chamber is half the length of a diagonal line drawn corner to corner on the floor. Using the Pythagorean theorem, it can be seen this has an accuracy of less than a quarter inch. Hence, the axial mode for the height and the diagonal tangential mode in room are directly related. This also creates an additive emphasis at a specific frequency, in this case approximately 58.8 to 60 Hz. In the same manner as the 2:1 ratio seen in the King's Chamber length and width, this ratio of height to diagonal of the floor creates another significant increase in the amplitude (energy) within the chamber. This frequency is very close to an octave of Danley's "about 30 Hz" and probably contributed to the phenomenon. There are other Pythagorean triangles in the King's chamber, but they involve oblique modes that are much weaker.

REGARDING DANLEY'S F# CHORD

Over the years, much has been written on Danley's comment regarding an "F-sharp chord" in the King's Chamber. I consider this to be a misunderstanding of his statement. As was mentioned earlier, I can see no good reason, nor any evidence, that the ancients used our contrived modern Western musical scale. This becomes obvious in the

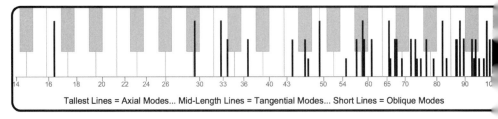

Tallest Lines = Axial Modes... Mid-Length Lines = Tangential Modes... Short Lines = Oblique Modes

Figure Aa.9. The dominant modes compared to a modern keyboard.

Grand Gallery corbels analysis and the foot stomp impulse measurements. Further, the modal analysis and computer simulations indicate frequencies that simply do not directly correlate to modern notes. The next chart superimposes the main resonant frequencies in the chamber over a modern keyboard.

A modern F# chord consists of the notes F#, A#, and C#. Using multiple computer modeling programs, frequencies corresponding to these precise notes are not to be found in the King's Chamber. For example, the models show an F# plus 2 to 27 cents in the chamber. Interestingly, I have measured this frequency in the Red Pyramid at Dahshur (F# plus 18 to 23 cents) as well. Similarly, the A# appears in the King's Chamber as A# plus 10 to 19 cents in the models. The C# is more problematic. It is missing in the lowest octave, a near miss in the second octave (C# minus 2 cents), and rather inconsistent for the next three octaves. Consequently, I suspect Danley's statement was intended as an illustrative generalization and other authors are taking it too literally.

THE UPPER CHAMBERS—THREE FUNDAMENTAL OBSERVATIONS

1. A pattern exists showing a direct relationship between resonant waves in the King's Chamber and the overall height of the Upper Chambers.
2. Another pattern exists showing a paired relationship of pressure zones that could facilitate vibrational movement of the blocks themselves.
3. Granite under stress can release significant amounts of energy.

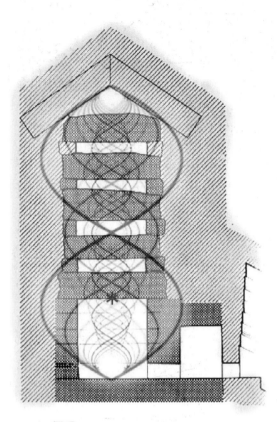

*Figure Aa.10.
Visualization of
harmonically related
frequencies at full
resonance in the
Upper Chambers.*

Figure Aa.10 shows the relationship of the height of the Upper Chambers, including the King's Chamber, to the height of the King's Chamber itself. There is a simple equation that illustrates this relationship: 0.5 (King's Chamber height) × 7 = height of the Upper Chambers complex.

The "× 7" in the above equation possibly indicates a preference for odd harmonics. The five blocks within the Upper Chambers are essentially flat on the bottom yet are very rough and irregular on the top. The flat surfaces would be more conducive to the creation and sustainability of standing waves. It is also possible the irregularities seen on the top of the blocks may have been a way to tune them. Removal of the material would be analogous to adding or subtracting water in a glass, thus changing its tone when struck. Since the blocks were never intended to be seen, it is puzzling why the ancient builders would do

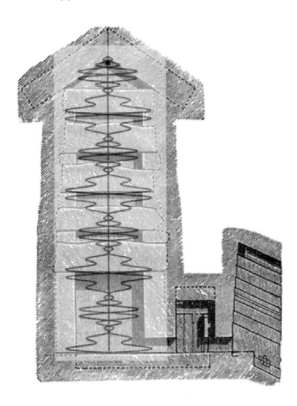

*Figure Aa.11.
Alternating high-
pressure and
low-pressure zones
in the Upper
Chambers.*

that much extra work unless it was necessary for some reason. This preference for odd harmonics is also seen in the Ascending Corridor. Due to the presence of the blocks in the lower end of the corridor, it is essentially a closed-end pipe that will produce resonant waves at the fundamental tone along with only odd harmonics.

Figure Aa.11 is a visualization composed of two elements. The waveform is a composite of a fundamental tone along with the 3rd, 5th, and 7th harmonics. That wave was then reflected back down to generate a waveform envelope that simulates what it would look like within the Upper Chambers. This produces a pattern of relatively higher and lower pressure zones. As can be seen, a standing wave composed of this harmonic structure aligns these zones in pairs, where a high-pressure zone above the block is countered with a lower-pressure zone below the block, or vice versa. This would probably enhance the ability of the blocks to flex or vibrate.

FINAL PIECE OF THE PUZZLE— GRANITE UNDER STRESS

Laboratory tests performed separately by Professor Mamoru Kato in Japan and Friedemann Freund at NASA have shown that granite, when properly stressed, will release enough energy to ionize the air surrounding the sample. It takes energy on the order of tens of thousands of volts per square inch to produce this effect. See color plate 5 (Kato 2007).

Freund's research repeatedly comments on the subtle forces necessary to release significant energy from within granite. Lab tests have shown forces roughly equivalent to a hand hit or foot stomp are enough to trigger the release of energy in granite. Since the breaking of bonds and the ensuing current flows all occur at quantum energy levels, the range of motion necessary to achieve this release of energy is measured in angstroms. A major variable to consider here is the stunning speed of sound in granite, approximately 13,309.8 miles/hour (5,950 meters/sec), as compared to a leisurely 767.3 miles/hour (343 meters/sec) for the speed of sound in air at normal ambient conditions (The Engineering Toolbox 2022). Since sound is a pressure wave, the faster the speed, the harder it impacts any object in its path. The effect of this dramatic difference here is worthy of contemplation.

Given that there are literally thousands of tons of granite in the Upper Chambers, one intriguing question remains to be tested: Can sound vibrations, properly tuned to the resonance of large granite blocks, actually vibrate the granite enough to release the energies? One key indication that this is possible could be the Ringing Obelisk at Karnak, where a simple hand hit causes a huge (70–80 tons or more) block of solid granite to ring like a bell.

CLOSING SUMMARY

This paper has explored areas in the pyramid that exhibit a rather unique and intriguing set of acoustic properties. Certain frequencies stand out as being produced, enhanced, or amplified by multiple aspects of the design and dimensions throughout the structure. Further, I believe this supports a conclusion that the pyramid is functioning as a system. To paraphrase the research of Chris Dunn, it appears to be a device, not just a collection of rooms and passageways. It is my sincere hope that other researchers interested in this line of thinking will find this commentary useful cannon fodder in their efforts and discussions. Finally, if serious acoustical testing of the pyramid will someday be done, answers to many of the questions posed throughout this paper may perhaps be known.

MAY 2022

ROBERT VAWTER graduated *cum laude* from San Jose State University with a degree in Archaeology & Anthropology in 1998.

Beginning in the mid-1970's he worked over twenty years as a sound engineer and lighting technician in small music venues and large auditoriums for local and national touring bands. Eventually this work expanded into the more analytical realms of sound in technical theatre and recording studio applications.

Combining his degree in archaeology with his expertise in how sound behaves in a wide variety of venues, Vawter was drawn to the study of ancient acoustics and the Pyramids of Egypt. Scholastic studies were augmented by on-site research in Egypt with Dr. Abd'El Hakim Awyan in 1996 and 1998.

Vawter's acoustic research in the Red Pyramid at Dahshur captured a range of frequencies that he used to create a composition titled, "The Dahshur Frequencies." A sought-after presenter at numerous conferences, including the National Convention for the Society for the Anthropology of Consciousness, he was also a frequent guest lecturer at San Jose State and Sacramento State Universities. In 2019, Vawter became a member of the GeoCosmo Research Group.

Robert Vawter passed away unexpectedly on September 17, 2023. He will be greatly missed.

APPENDIX B
USING SEMICONDUCTOR PHYSICS TO FORECAST EARTHQUAKES

FRIEDEMANN FREUND, PhD

I came to Christchurch to talk about earthquakes. There's no better place in the world, I would say, to talk about earthquakes, and particularly, I want to give you hope. So, if you look at the photo of the New Zealand islands, it looks so peaceful and classic—you see Christchurch on the left side. But on certain days, occasionally, the Earth speaks to us with a very violent voice, and many of you have gone through this experience. I would like to present you with a question. What if we would be able to see these earthquakes coming days before they arrive?

Now, I worked for the past 30 years for NASA in California. Before this I was a professor in Germany, and during that time I was doing things that at first had no relationship to earthquakes. I was studying single crystals with my son, Nino, who was a director of nanotechnology

---

*Transcribed from a TEDx lecture presented by Freund in Christchurch, New Zealand, on December 12, 2016. Transcription and images copyright of Friedemann Freund and used here with his permission (Freund 2016).

289

and advanced space science at NASA Ames—and he passed away not long ago from a brain tumor. So much of what I am saying is in honor of my cooperation with him.

I started out working on very simple single crystals, totally, as I said, unrelated, never dreaming that what I was working on would one day have influence and importance for understanding earthquakes.

Magnesium Oxide. The simplest oxide material that exists, and I worked on it for a number of years, and I discovered something that everybody else had overlooked—had missed. And that is a defect. A type of defect in these crystals. And I should say that a defect in our world of solid-state physics is something that appears not as a crack, it is something that appears on the atomic, on the subatomic, realm. What I discovered was that in the crystal structure, like here, on the lower left side, there are these defects that are totally invisible. See color plate 2.

Until now we have no way to directly observe this peroxy defect, but when we do nasty things to the crystal they fall apart and produce electric charge carriers. Then I found the same kind of defects in other materials, including natural minerals either from the crust or from the Earth's mantle. Almost every mineral seems to have these, and if minerals have them, then, of course, rocks will have them.

So now I want to show you what you can do when you play around with rocks. So here you see a four-meter-long piece of granite in my laboratory at NASA Ames, and all what I am doing here is I put some contacts to the rock at the far end and the contact up here and I squeeze here. In the moment I start squeezing here, a current starts to flow through the unstressed rock. And if I put an ammeter in the circuit, I can measure an electric current. And here you see this example measure from this particular four-meter-long piece of granite. The green curve is the load. The moment I start to load the rock, the current begins to flow. The current rises very rapidly at low stress levels, saturates, and stays constant; actually—we have measured this over months—the current continues to flow if we keep it loaded. And if we unload it, the current goes away. See color plate 3.

Now those of you who have maybe heard about how semiconductors work, this is a behavior that is reminiscent of a semiconductor—

the thing from which you build transistors and that becomes available in all electronics. Here at the bottom you see the schematic. On the right-hand side, the stress activates electrons and holes. Holes is the name for the defect and electrons are electronic particles that are generated and are necessary to be able to produce a thermistor. And we found out that electrons have to stay in the stressed rock volume while the holes flow out. They flow out at about 100 meters per second, which is about the speed of a jet landing at an airport. They are very fast. They propagate through it, and in this particular experiment, the electrons have to come around through the wire and shake hands at the front end of this rock.

Now this is a combination of a semiconductor behavior of electrons and holes and an electrochemical battery, which means that we can separate the positive holes flowing through the rock from the electrons that flow through the wire. This is exciting and this is new. Now let's do the Duncan experiment. Take this rock and imagine that it would be sitting vertically in the Earth's crust and that this rock is not four meters long but a kilometer, five, ten, thirty, or fifty to a hundred kilometers long if we are in a subduction zone like we are in the North Island. See color plate 4.

If that end of that volume of rock is stretched out by the enormous tectonic force in the Earth, electricity is generated, and under certain conditions, electricity flows out of the stressed rock volume and into the surrounding rock. We have means to see these things flowing over distances of tens of kilometers under natural conditions. These are amazing properties that have never before been properly understood.

Now if these holes—because they have such unusual behavior—they have been given the name positive holes. I didn't give it; a good friend of mine gave it in the 1990s. If a current flows through something, it generates a magnetic field. If this current is constant and strong enough, even if it flows at a ten- or twenty- or thirty-kilometer depth, we can measure the change in the magnetic field at the Earth's surface. If the current fluctuates, which it most often does, then it becomes antennas emitting radio waves, and the ultra-low-frequency portion of these radio waves can travel through tens of kilometers of rock and we can measure it. And then you see here unipolar pulses, and these are amazing. I will show you an example afterwards.

If they start to flow into water, chemistry takes place. Electrochemistry. The water becomes stoichiometrically oxidized to hydrogen peroxide, the stuff you put on your skin when you have a scratch. You change the chemistry of water and you change the cation in iron, and iron content you change the fluorescence spectrum of the dissolved organic that is dissolved in all groundwater and river waters.

Then there are ground potential building up that we can measure. When we see this ground potential building up, if they reach a certain threshold, we start to see ionization of the air. First, positive ions are formed over an area about a hundred or two hundred kilometers wide; suddenly the air becomes full of positive ions. At the same time, we see carbon monoxide seeping out of the ground. If the Earth continues to stress it, more and more charges come; we start triggering corona discharges. In the lab, we can photograph these little corona discharges.

Out in nature, they instantly form negative air ions that we can measure, the negative air ions that form ozone and other light products, and on the left side, you see there the infrared emission, and I will also show you an example of this. Then there are many, many perturbations in the uppermost atmosphere—a hundred to two or three hundred kilometers above the ground which have been reported extensively in the literature. Very few people, I would almost say nobody really, understood how these things are being generated. So now all these things we can now use to learn something about the stress state of rocks, deep below our feet. Far beyond the direct reach, we have to deduce their presence. From these indirect measurements we can establish these causality ranges linked together by chemistry and by physics.

So, I wanted to briefly mention this infrared emission. The infrared emission is when the charge carriers come to the surface and can recombine. During this recombination, these charge carriers release energy, and these two oxygens, they become suddenly about 20,000 degrees hot vibrationally and they emit bursts of infrared radiation. But one characteristic feature is that these charge carriers try to accumulate mostly on mountain tops. On hillsides, not in the valleys. See Figure Ab.1 and color plate 5.

Figure Ab.1. Granite under stress generating enough energy to ionize air. (See also color plate 5.)

Now I want to show you here the result from a PhD thesis by Luca Piroddi in Italy, who had analyzed, I think, 18 months of Italy. Every night he looked for these anomalies, and prior to this earthquake in L'Aquila in 2009, he identified this anomaly that you see here in red. And here's a photograph of L'Aquila, a medieval town that was heavily damaged with a loss of life of more than 300 people, and to the left and to the right there are mountain ranges. The ones at Gran Sasso are the highest mountains in the middle of Italy. In the next slide you will see draped over exactly the same, this map of the thermal infrared anomalies as they are measured, three days before that disastrous earthquake. So, if anybody would have had the funding and the knowledge to analyze how important the analysis of this phenomena in real time three days before the earthquake was, they would have issued a warning and say something is brewing. Something must be happening 10 to 20 kilometers (6.2–12.4 miles) below our feet.

Now, next I want to talk about these unipolar pulses. Unipolar pulses are strange phenomena that, suddenly, within a fraction of a second, enormous energy—electromagnetic energy—is emitted from deep below. Only about 100, 150, 200 milliseconds long, [the pulses] shoot up, come down, and with a little wiggle, disappear. They are not yet fully understood; we are working on it.

Now here, the situation where a friend of mine, colleague Jose Haro, has operated for the past two-and-a-half years a station consisting of two search coil magnetometers, extremely sensitive, and what you see there is, in the subduction zone there is a submarine ridge where earthquakes are being generated. That ridge is subducted and disappears underneath the edge of the South American continent.

Now here is again the map, and you will see in a moment how these unipolar pulses are generated, and they are marked here. This is a period of about two weeks that you see, and here they come. A few 10 minutes distance from each other impact in groups. Then there is a day or day and a half of silence, no unipolar pulse, then the next group is coming again a little silent, or there is another blip from somewhere far away, and then there is a third group of these pulses coming. Now we have 2,500, more than 2,500 of these pulses, and they were associated with 22 magnitude-4 earthquakes, 3.5 to 4 earthquakes that happened in this subduction zone between the depths of 25 to 65 kilometers.

Jose Haro is able to predict a time window of 40 hours, three to six days in advance, depositing the information in a closed envelope with the president of his university and in all 22 cases that he has analyzed, he was right. It was quite remarkable.

So, I've started my presentation with asking "What if we can see earthquakes coming?" I think I can say I hope I have convinced you that by understanding the physics—and this is really new physics of how rocks respond deep below to the increase of stress, and how/what they are producing electricity, and how this electricity propagates through the Earth's crust—we can learn about the buildup of stress and we will be able to say, "Yes. We can see them coming, days before they arrive." We cannot be sure that every stress buildup will lead up to an earthquake because sometimes the Earth says, "I don't feel like rupturing, I feel like sliding." Then we would see the precursors and people would say that you have had a false alarm. No, we have not had a false alarm. The Earth has just had another day. Another feeling. But in most cases, one thing we can make for sure is that in the future there will be no major earthquake that will hit any place in the world where we have the capability of measuring these precursory signals, will hit unannounced, unforewarned. So, the element of utter surprise that has been a plague until recently, and including the Christchurch earthquakes, will be over. Thank you so much.

FRIEDEMANN FREUND holds a PhD in Mineralogy and Crystalography from Marburg University, Germany, 1959.

Professionally, Freund is a Principal Investigator at the SETI Institute (since 1989), an adjunct professor in the physics department of San Jose State University (since 1989), as well as a NASA Associate at NASA Ames Research Center (since 1985).

Freund has published 300 papers in peer review journals. His research interests include defects in crystals; proton conductivity; defects in minerals caused by the incorporation of H_2O, CO_2 and other gas/fluid components; prebiotic chemistry in the solid state and origin of life; valence fluctuations in the oxygen anion sublattice in oxide/silicate minerals; and rock physics in relation to earthquake and pre-earthquake phenomena.

In 2014 he founded the GeoCosmo Research Centre, an independent think-tank comprised of scientists from around the world. A more complete CV of Freund's work can be viewed at the SETI Institute website.

APPENDIX C
LARGE-SCALE THERMAL ACOUSTIC GENERATOR

ERIC E. WILSON

ABSTRACT

This paper discusses a method of combining a man-made system with a unique natural phenomenon to generate power on a very large scale by harnessing a natural reaction common to the earth. This energy system taps into a very unique electrical phenomenon for clean and nearly free energy. Natural energy production is witnessed all around us. On (or within) the earth there are many types of natural electrical phenomena. Most are recognized in the electrical coupling between atmospheric charges and the earth manifesting as lightning. However, the most intriguing fact is that electricity is found at ground level, generated by the properties of certain types of rock in very large formations. Electricity has been observed and generated from stone if the stone is vibrated at the correct infrasonic frequency range. A thermoacoustic

*As presented to the American Institute of Aeronautics and Astronautics (AIAA) 2019 Propulsion and Energy Forum. 19–22 August, 2019, Indianapolis, IN. Edited with permission.

engine reaction matches the optimum frequency of power generation from stone.

I. NOMENCLATURE

Abbreviation	Meaning
eV	electron volt (energy)
Hz	hertz

II. INTRODUCTION

The purpose is to present a combined concept. No technical data will be shared in this document. Evidence is presented from satellite observations and lab testing showing that the coupling of very large natural rock formations, or quarried rock, or stacked stone, can generate electrical currents and fields when exposed to compressional stresses or vibration. The induction of the vibration is proposed using a phenomenon of a well-published device known as a thermoacoustic engine. The combination of these two systems results in a high-voltage production of power within rock. These charges migrate to the surface and subsequently couple with the atmosphere and ionize the surrounding air. The reaction can be harnessed as a method of power generation.

III. THE THERMOACOUSTIC ENGINE

Study of thermoacoustic engine design has shown that these engines, when properly tuned, are capable of indefinite output with no active moving parts when provided minimal heat input into the hot section. A nonlinear response has been found that makes the engine purposefully unstable. The instability is exploited for the vibration, which is useful in energy production. The thermoacoustic engine reaction "stack" optimally resonates in a range of 3–6 Hz. This reaction frequency is very important as it matches the study we will reference in section IV.

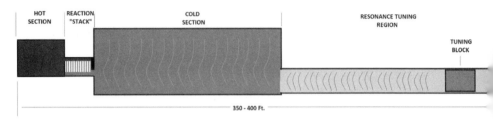

Figure Ac.1. Example scale of the large-scale thermoacoustic structure.
Appendix C images credit: Eric Wilson.

The structure of the thermoacoustic engine proposed in this paper for large scale is depicted in Figure Ac.1.

The referenced work on thermoacoustic engines is credited to Purdue Propulsion. A published paper from the 2018 AIAA Fluid Dynamics Conference introduces a response that was revolutionary and somewhat unexpected as it was a nonlinear exponential response (Alexander et al. 2018).

Testing is being done on a small scale and on individual subsystems of the proposed system. No proposition has been made except in this paper for a very large-scale system. Reference is given from published work at Purdue University, where graduate students are working out the optimization of thermoacoustic engines. Details of that study will reinforce optimum tuning of these devices. With these experimental engines, scalability is being determined. Some proposed sizes are being discarded and considered as impractical for their task at hand because of the targeted pursuit of smaller machines to work with heat recovery from gas turbines. In contrast, larger-scale machines are considered in this paper for practicality in large-scale power generation (Alexander et al. 2018).

IV. FREUND EFFECT—ELECTRICITY IN ROCKS

The work of NASA's Friedemann Freund, PhD, has linked ground-level electrical anomalies before and during earthquakes, in specific types of rock formations, to what have been labeled in geophysics as earthquake lights. This phenomenon is known by those in Freund's close circle of researchers as the Freund Effect (Freund 2003, 2000).

Due to response levels with voltages high enough for ionization, the system will avoid traditional conductors to extract power from the reaction core. Instead, we will incorporate waveguides filled with a pure ionized gas excited to a plasma state. When gasses are excited to their plasma state, they perform as very high-energy capacity superconductors.

The dream of many to harness energy from sources such as lightning has not yet been achievable. We are not exactly harnessing lightning, but it is a good analogy.

More and more, examples in nature have been used in the design of advanced technologies. For example, Apple has exploited this engineering tactic in the discovery of their fractal antenna. This antenna mimicked the natural random structure of a tree to achieve an antenna with superior reception.

A. Solid State Processes

There are two main elements to the theory of earthquake lights. Those are acoustic vibrations and electrical charges. Stress and vibrations in igneous rock create electrical charges. This electrical phenomenon is not well-known outside of earthquake lights research. One minor part of the reaction is the piezo effect, but it is not the primary driver of electricity generation in vibrating rock.

Granite has the best observable reactions. In the lab example in Figure Ab.1 and color plate 5, granite under stress generates electrical charges sufficient to ionize air before failure in the sample. Oxygen atoms reacting produce 630 nm photons in the 1.97 eV range. As the rock is heavily stressed, in this case to a breaking point for highest output, ionic bonds in quantum valence levels break and then recombine. There is a releasing of energy in this process, and the charges migrate to the surface (Freund 2009).

B. Peroxy Anion Effects in Stone

When granite is formed by heat and cooled in the presence of water, hydroxyl bonds are formed (Freund et al. 2021).

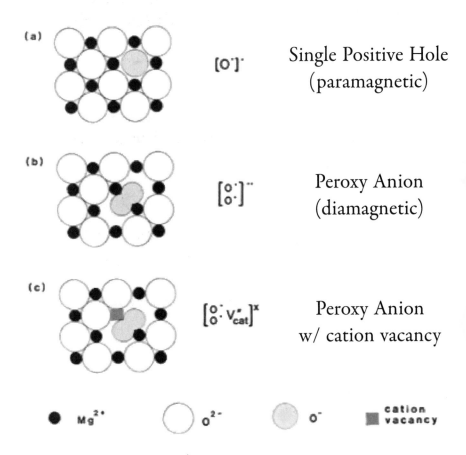

Figure Ac.2. Peroxy anions form in granite. (Bruce V. King and Friedmann Freund 1983)

C. Seismic Vibration Response

When seismic activity disturbs the hydroxyl bonds, electrons seek to form new bonds (Freund et al. 2021).

D. Power Generation

The schematic in figure Ac.4 depicts the testing apparatus used by Freund and the propagation of ionization (Freund 2000).

Normal/Quiet
Bonds Are Stable

Seismic Activity
Bonds Break Apart

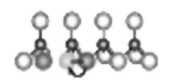

Electrons Seek to
Reform New Bonds

Figure Ac.3. Hydroxyl bonds seek to form new bonds.

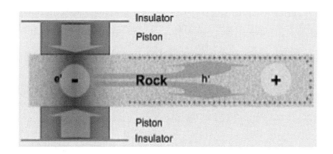

Charge
Generation
and Flow

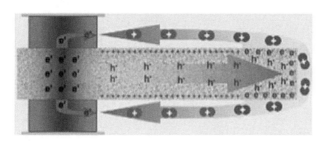

Field
Generation
and Air
Ionization

Figure Ac.4. Freund's test rig propagation schematic.

V. ACOUSTIC COUPLING

The thermoacoustic engine reaction optimally occurs in a range of 3–6 Hz. This reaction is also optimum for the Freund Effect reactions measured in igneous rocks, such as basalt, gabbro, granite, and certain types of feldspars, for producing electricity. Earthquake infrasonic signatures are in this frequency band when at peak activity and are synchronous with the measurement of these energy production reactions recorded by ground and satellite studies (Piroddi et al. 2014; Aizawa 2004).

Amplification can be achieved in the production of energy when the target rock is machined and tuned. A stack of stone subjected to compression from a mass structure will exhibit the greatest output. The stones can be tuned to a harmonic of the proposed acoustic engine well above 3–6 Hz range. Resonance chambers are used to produce the desired frequency environment as long as there is a harmonic link to the source reaction frequency. The primary system overall has no moving parts.

The Freund Effect has measured the energy being transmitted over 100 miles and migrating to the highest point in topographical formations of rock. The potential difference has been estimated in natural production up to 100 Kv and up to 6M amps. It is conceivable that an array of similar tuned devices could be placed in appropriate locations and produce power in tune with the primary resonant seismic energy source (Thériault et al. 2014; Uyeshima 2007).

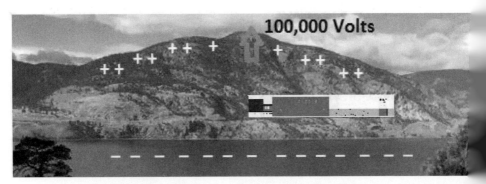

Figure Ac.5. Power is generated by vibration and large-scale ionization.

VI. CONCLUSION

The largest proposed thermoacoustic machines are tuned with the combined length of all chambers approximately 350–400 feet (106.68–121.92 meters) in length for best performance, narrow and long, much like a subway underground station or a bored tunnel. The construction would be best served inside of a mountain.

The power plant using these proven subsystem technologies would benefit when placed in a location of natural rock and structured with acoustic chambers of stone tightly fit without mortar. The best location for this structure would be near a flowing river or the dam discharge flume from a lake or a pond.

Thermoacoustic engines sometimes require a hammer-force shock to start the reaction. The force of a natural water flow can be used in conjunction with the concept of a hydraulic ram pump to supply this force inside the structure and to supply cooling for the reactions inside the structure. The hydraulic ram pump driven on a large scale is capable of producing several tons of shock force. The magnitude of this force would be sufficient to shock the thermoacoustic system into action. Once the hydraulic ram triggers the first vibrational shock to start the system, the reaction will continue until the process is mechanically interrupted.

ACKNOWLEDGMENTS

Special thanks to Friedemann Freund and the NASA GeoCosmo Research Centre.

Eric Wilson: Technical Specialist—Thermal Management, Research and Technology.

ERIC WILSON received an undergraduate degree in Advanced Technical Studies from Southern Illinois University in 1997, combining an aviation, electrical, and mechanical engineering degree.

In 2008, Wilson graduated with a Master of Technology from Purdue University in Lafayette, Indiana.

Currently, Wilson is based in Giza, Egypt as a technical specialist in thermal management, electrical machines, and cyber security for a major gas turbine engine manufacturer.

Wilson fills his time in Giza by continuing to research ancient technology, and since 2019 has provided volunteer assistance to advance pyramid scanning projects.

APPENDIX D

THE GREAT PYRAMID OF GIZA AND THE QUEEN'S CHAMBER SHAFTS

A CHEMICAL PERSPECTIVE

Brett I. Cohen, PhD

ABSTRACT

In his book *The Giza Power Plant: Technologies of Ancient Egypt* (Dunn 1998), Christopher Dunn proposes a very intriguing theory regarding the function of the Queen's Chamber. He suggests that it was an engine of a great power plant, where hydrogen (H_2) gas was used to create piezoelectric or electromagnetic energy. A detailed chemical approach has been applied to the chemicals used in the southern and northern shafts of the Queen's Chamber. In the South Shaft of the Queen's Chamber, sulfuric acid (H_2SO_4) was used, not hydrochloric acid (HCl) as Dunn describes. In the northern shaft of the Queen's Chamber, a salt

*Published in *Nexus* August–September 2015. Reprinted here with the kind permission of the author, and Duncan Roads, Publisher. © Brett I. Cohen, PhD, 2015

mixture of ammonium chloride (NH_4Cl) and zinc chloride ($ZnCl_2$) was used. Hydrogen gas could be generated from both the southern and northern shafts using the chemicals described. A mixing of the southern and northern shaft chemicals was not necessary for the production of H_2 gas. Each system described in this paper could be a stand-alone source of H_2 gas.

INTRODUCTION

The Great Pyramid of Giza is the oldest and largest of the three pyramids on the Giza Plateau (or Necropolis) in Giza, Egypt. Markings inside chambers in the Great Pyramid reference the construction of the pyramid in the Fourth Dynasty of Egyptian Pharaoh Khufu. The Great Pyramid is the oldest of the Seven Wonders of the Ancient World, the only one that remains largely intact. It is also known as the Pyramid of Khufu or the Pyramid of Cheops. It is believed by Egyptologists that the pyramid was built over a ten- to twenty-year period and that construction was concluded in about 2560 BC (Wikipedia 2001).

In *The Giza Power Plant*, Dunn proposes that the Queen's Chamber in the Great Pyramid of Giza was an engine of a great power plant. Here is a brief description of Dunn's theory.

1. Each of the shafts (southern and northern) found in the Queen's Chamber held separate chemicals, the South Shaft containing dilute hydrochloric acid and the North Shaft having hydrated zinc chloride.
2. When these chemicals were mixed together, hydrogen gas was produced.
3. H_2 gas filled the Grand Gallery and travelled up into the King's Chamber, where acoustic energy was converted to piezoelectric energy.

In this paper, each of the shafts (southern and northern) of the Queen's Chamber and their chemicals as Dunn describes are analyzed for chemical validity, and chemical alternatives are also introduced and discussed. Only the chemistry regarding the Queen's Chamber shafts (southern and northern) and its production of hydrogen gas is discussed.

USE OF SULFURIC ACID (NOT HYDROCHLORIC ACID) IN THE SOUTHERN SHAFT OF THE QUEEN'S CHAMBER

It is proposed that sulfuric acid (molecular weight [MW] of 98.08 g/mol) was used in the southern shaft of the Queen's Chamber. H_2SO_4 is a strong, highly corrosive mineral acid and its historical name is oil of vitriol, the archaic name for sulfate (Karpenko and Norris 2002). Dunn describes the use of dilute hydrochloric acid (HCl) in the southern shaft. He also explains that gypsum residue was found in this shaft. Gypsum is a soft sulfate mineral (calcium sulfate dihydrate, $CaSO_4 \cdot 2H_2O$). Dunn states that the limestone (a sedimentary rock composed largely of the mineral calcite, $CaCO_3$) surrounding the shaft would undergo a chemical reaction with HCl to produce gypsum. However, HCl cannot produce a gypsum adduct since no sulfate (SO_4) is found in its formula, therefore H_2SO_4 is the likely acid used. The chemical substitution reaction below illustrates the production of gypsum adduct from H_2SO_4:

$$H_2SO_4 + CaCO_3 \text{ (limestone)} \rightarrow \begin{array}{l} CaSO_4 \text{ (gypsum)} \\ + CO_2 \text{ (carbon dioxide)} \\ + H_2O \text{ (water)} \end{array}$$

The source of H_2 found in the southern shaft could be explained by the dissociation of aqueous H_2SO_4. The use of vitriols dates back to antiquity and is recorded in the Sumerians' word lists (dictionaries) written c. 600 BC (Crosland 1962). In these word lists, all types of vitriol are described according to color. The green vitriol is iron(II) sulfate with seven hydrated water molecules ($FeSO_4 \cdot 7H_2O$) (Crosland 1962). In ancient times, green vitriol was called copperas.

A survey of the chemical literature has not found any reference that the ancient Egyptians used H_2SO_4, but here it is shown that they had the technology and resources to produce large quantities of this acid. Sulfuric acid is a strong, highly corrosive acid. The production and storage of H_2SO_4 would have been problematic for the ancient Egyptians.

The most abundant artifact found in Egyptian archaeological sites is ceramic clay pottery. The ancient Egyptians fired marl clay ceramic pottery at very high temperatures (between 800°C and 1000°C), and these ceramic materials were very hard after firing (McGovern 1997).

The marl clay ceramic pottery usually did not contain any organic materials after firing. Such a ceramic vessel would have been ideal for the production and storage of H_2SO_4 due to its high chemical resistance and strength. The mineral known as green vitriol would have been used for the production of H_2SO_4 (Schröder 1957). The chemical procedure would be to place the solid green vitriol mineral in a marl clay ceramic vessel and heat it to approximately 1000°C for a few hours. The resulting liquid produced, H_2SO_4, would be decanted off and placed in another marl clay ceramic vessel for storage. The resultant solid residue of green vitriol would then be discarded and the process would be repeated numerous times using new green vitriol mineral.

This chemical procedure is called a dry distillation, and such a procedure is found in the chemical literature.

In 1957, a German chemist, Gerald Schröder, reported a dry distillation procedure using green vitriol. Schröder's procedure is as follows. A quantity of 200 grams (0.72 mol) (approximately half a pound) of green vitriol ($FeSO_4 \cdot 7H_2O$, MW of 277.85 g/mol) was gradually elevated in temperature to 1000°C over a three-hour period, after which time approximately 8.0 grams of a "strongly acidic liquid, smelling like SO_2," was obtained.

Upon chemical analysis, this resultant liquid had approximately 11% H_2SO_4 (Schröder 1957). (If the reaction had gone to completion, a total of 70.60 grams of H_2SO_4 would have been produced. Here, the percentage yield is 8.0 grams / 70.60 grams × 100% = 11.33%.)

So, in a very simple procedure, 200 grams of starting material (green vitriol) was heated, and this resulted in the production of approximately 8.0 grams (0.082 mol) of H_2SO_4.

The ancient Egyptians would have repeated this process over and over again to produce large quantities of H_2SO_4. They had the means of achieving the required temperatures from the technology of producing marl clay ceramics, and these ceramic vessels would have been ideal for the production and storage of H_2SO_4.

All different shapes and sizes of marl clay ceramic pottery have been found in archaeological sites. Ceramics with large-diameter lower-vase shapes with small-diameter mouth and neck openings have also been found (McGovern 1997).

This shape would have been ideal for the production of H_2SO_4 because it would have reduced exposure to the fumes of sulfur dioxide (SO_2) gas and hydrogen (H_2) gas in the heating process. Sulfur dioxide gas is a toxic gas with a pungent, irritating, and rotten smell. Green vitriol allows the dry distillation process to occur by virtue of the seven hydrated water molecules found in the mineral compound. When the mineral is heated as described, the solution obtained is a water-based solution of H_2SO_4.

It should be noted that modern chemical processes (chemical reactions in aqueous solution, not dry as the one proposed for the ancient Egyptians) for the production of H_2SO_4 are much more efficient than the procedure described above.

The ancient Egyptians would have put more manpower into the "dry distillation" process and, over time, they would have produced significant quantities of H_2SO_4.

USE OF AN AMMONIUM CHLORIDE–ZINC CHLORIDE SOLUTION IN THE NORTHERN SHAFT OF THE QUEEN'S CHAMBER

Dunn also describes the use of a hydrated zinc chloride ($ZnCl_2$) solution in the northern shaft of the Queen's Chamber. $ZnCl_2$ is a very soluble solid in water, and in the solid form it is extremely hygroscopic (it absorbs water very quickly from the surrounding air).

At the very end of the northern shaft, a limestone door with two metal handles was found, one handle appearing to have a white coating on it and surrounding it.

Dunn believes this coating was formed by some type of electroplating process.

If this is true, it is proposed here that a solution of ammonium chloride (NH_4Cl) would also have been needed for the electroplating process to occur without extreme heat. NH_4Cl is a white, crystalline, inorganic salt compound that is extremely soluble in water. Aqueous solutions of NH_4Cl are mildly acidic (see EQ1).

A form of electroplating (also called galvanization or hot-dip galvanization) exists where a metal (iron, steel, or copper) is placed in a bath of molten zinc (Zn) at a high temperature (approximately 500°C) (Strutzenberger and Faderl 1998).

This process would have been problematic for the ancient Egyptians, so, in order to achieve the electroplating effect that Dunn describes, they needed to use another chemical, NH_4Cl.

A type of electroplating called electrogalvanization exists, where a thin layer of zinc is deposited and bonded to metal (where a controlled thickness of Zn may be applied and deposited) by running a current of electricity through a salt (saline)–zinc solution, and this process can be performed at lower temperatures, such as at room temperature (Chandrasekar and Pushpavanam 2008).

It is proposed here, where multiple simultaneous reactions are occurring at room temperature, that with the use of an electric current from the production of piezoelectric energy from the complete Giza power plant process (see Introduction), a solution of NH_4Cl (salt) and $ZnCl_2$ could achieve the electroplating effect. The reactions are noted in the equations below.

In EQ1, aqueous NH_4Cl results in the dissociation into NH_3 gas and HCl.

In EQ2, aqueous $ZnCl_2$ and HCl results in the production of metallic Zn for the electroplating process, an unstable zinc hydride (ZnH_2) and chlorine (Cl_2) gas.

In EQ3, ZnH_2 results in the production of metallic Zn for the electroplating process and H_2 gas. Aqueous ZnH_2 has been observed to decompose rapidly in acidic solutions (HCl found in EQ1 and EQ2) (Finholt et al. 1947) to metallic Zn (used for the electroplating process) and H_2 at room temperature.

EQ1
$$NH_4Cl \text{ (ammonium chloride)} \rightarrow NH_3 \text{ (ammonia gas)} + HCl \text{ (hydrochloric acid)}$$

EQ2
$$2HCl + 4ZnCl_2 \text{ (zinc chloride)} \rightarrow 3Zn \text{ (metallic zinc)} + ZnH_2 \text{ (zinc hydride)} + 5Cl_2 \text{ (chlorine gas)}$$

EQ3
$$ZnH_2 \text{ (zinc hydride)} \rightarrow Zn \text{ (metallic zinc)} + H_2 \text{ (hydrogen gas)}$$

Some of the gasses produced in EQ1, EQ2, and EQ3 (NH_3, Cl_2, and H_2) could be dissolved in the aqueous salt (saline) solution of NH_4Cl and $ZnCl_2$. The formation of H_2 gas in EQ3 is a source of H_2 gas found in the northern shaft. Sal ammoniac is the name of the natural mineralogical form of NH_4Cl, and the mineral is commonly found on burning coal dumps (due to the condensation of coal-derived gas). Interestingly, the word "ammonia" (NH_3) is related to the classical-era discovery of sal ammoniac near the temple of Zeus Ammon in the Siwa Oasis in the Libyan Desert.

The ancient Egyptians were very familiar with NH_4Cl. At one stage in their history, they worshipped Amun (or Amon, a major Egyptian deity) and used NH_4Cl in their religious worship rituals. Also, several words are derived from Amun via the Greek form, Ammon, such as "ammonia" and "ammonite" (Sutton et al. 2008).

CONCLUSION

In conclusion, it is proposed that in the southern shaft of the Queen's Chamber, H_2SO_4 was the acid used, not HCl. In the northern shaft of the Queen's Chamber, a salt (saline) mixture of NH_4Cl and $ZnCl_2$ was used. It has been illustrated that H_2 gas could be generated from both the southern (H_2SO_4) and northern (NH_4Cl–$ZnCl_2$) shafts using the chemicals described in this paper.

It would not be feasible for the mixing of the chemicals of the southern and northern shafts to be efficient, since Dunn describes no mixing apparatus as existing in the Queen's Chamber.

Therefore, it is also proposed that a mixing of chemicals of the southern and northern shafts was not necessary for the production of H2 gas. Each system described in this paper could be a stand-alone source of H2 gas.

BRETT I. COHEN holds a PhD in inorganic and bioinorganic chemistry from the State University of New York at Albany. He received his PhD in November 1987 for his thesis titled "Chemical Model Systems for Dioxygen-Activating Copper Proteins" and was a postdoctoral fellow at Rutgers University in 1988–1989.

His research at Rutgers was in the area of peptide synthesis utilizing transition metal chemistry. After his postdoctoral fellowship, from 1989 to 2003, Cohen was one of the owners of Essential Dental Systems (manufacturer of dental composites and dental materials), where he was chief executive officer and vice president of dental research.

Cohen has had over 100 papers published in peer-reviewed journals (such as *Journal of the American Chemical Society, Inorganic Chemistry, Journal of Dental Research, Journal of Prosthetic Dentistry, Journal of Endodontics and Autism*, etc.) and has obtained 16 US patents.

His papers cover a variety of areas such as inorganic and bioinorganic chemistry, biomedicine, autism, physical chemistry, dentistry, and more. Cohen can be reached via email at ebicbis@aol.com.

BIBLIOGRAPHY

Academo. n.d. "Wave Interference and Beat Frequency." Academo website.

Adly, Ahmed. 2021. "Measures of the Box in Serapeum." Ahmed Adly YouTube channel, February 28, 2020. Accessed April 29, 2022.

Aizawa, K. "A Large Self-Potential Anomaly and Its Changes on the Quiet Mt. Fuji, Japan." *Geophysical Research Letters* 31, L05612 (2004).

Alexander, D., M. T. Migliorino, S. Heister, C. Scalo. "Numerical and Experimental Analysis of a Transcritical Thermoacoustic Prototype." Paper presented at the AIAA Fluid Dynamics Conference, Atlanta, Georgia, 2018.

Allen, C. "The Role of Animal Behavior in the Chinese Earthquake Prediction Program." In *Abnormal Animal Behavior Prior to Earthquakes*, J.F. Evernden (ed.), National Earthquake Hazards Reduction Program, USGS, Menlo Park, CA, 23–24 September 1976, 5–13.

Babbage, Charles, Philip Morrison, and Emily Morrison. 2012. *Charles Babbage: On the Principles and Development of the Calculator and Other Seminal Writings*. Mineola, NY: Dover Publications.

Balezin, Mikhail, Kseniia V. Baryshnikova, Polina Kapitanova, and Andrey B. Evlyukhin. "Electromagnetic Properties of the Great Pyramid: First Multipole Resonances and Energy Concentration." *Journal of Applied Physics*, 124, 034903 (2018).

Barsanti, Alessandro. 1906. "Fouilles de Zaouiét el-Aryán (1904–1905)." *Annales du service des antiquités de l'Égypte—Súppleménts (ASAE)*, 7: 257–287. Institut Français d'Archéologie Orientale, Kairo.

BrainyQuote website. n.d. "Arthur C. Clarke Quotes." Accessed April 9, 2022.

Bressan, David. 2020. "Nikola Tesla's Earthquake Machine." *Forbes* online, January 7, 2020..

Brown, David Jay, and Rupert Sheldrake. n.d. "Unusual Animal Behavior Prior to Earthquakes: A Survey in North-West California." Animals and Earthquakes Survey website. Accessed April 18, 2022.

Cayce, Edgar Evans, Gail Cayce Schwartzer, and Douglas G. Richards. 1998. *Mysteries of Atlantis Revisited*. New York: Harper Collins.

Chandrasekar, M. S. and M. Pushpavanam. 2008. "Pulse and Pulse Reverse Plating—Conceptual, Advantages and Applications." *Electrochimica Acta* 53 (8): 3313–3322.

Clarke, Arthur C. 1964. *Profiles of the Future: An Inquiry into the Limits of the Possible*. New York: Bantam Books. First published 1962.

———. 1968. "Clarke's Third Law on UFOs." *Science* 159 (3812) (1968-01-19): 255.

Clow, Barbara Hand. 2011. *Awakening the Planetary Mind*. Rochester, VT: Bear & Company.Cohen, Ariel. 2019. "Is Fusion Power within Our Grasp?" *Forbes* website, January 14, 2019.

Cohen, Brett. 2015. "The Great Pyramid of Giza and the Queen's Chamber Shafts: A Chemical Perspective." *Nexus* (August–September 2015): 43–46.

Crosland, M. P. 1962. *Historical Studies in the Language of Chemistry*. London: Heinemann.

Daniels, Jeff. 2017. "Mini-Nukes and Mosquito-Like Robot Weapons Being Primed for Future Warfare." CNBC website, March 17, 2017.

Danley, Thomas. 2000. "The Great Pyramid: Early Reflections & Ancient Echoes." *Live Sound International,* July/August 2000. Reprinted August 4, 2020, at ProSoundWeb Website. Accessed April 28, 2022.

de Bourrienne, Louis Antoine Fauvelet. 1831. *Private Memoirs of Napoleon Bonaparte, During the Periods of the Directory, the Consulate, and the Empire, Vol. I.* New York: Nabu Press.

Descartes, René. 2008. *Meditations on the First Philosophy: With Selection from the Objections and Replies*. Translated by Michael Moriarty. New York: Oxford University Press.

Drexler, K. Eric. 1987. *Engines of Creation: The Coming Era of Nanotechnology*. New York: Anchor Books. (Originally published in hardcover on January 1, 1986).

Dunn, Christopher. 1984. "Advanced Machining in Ancient Egypt?" *Analog Science Fiction and Fact*, August 1984.

———. 1998. *The Giza Power Plant: Technologies of Ancient Egypt*. Rochester, VT: Bear & Company.

———. 2010. *Lost Technologies of Ancient Egypt: Advanced Engineering in the Temples of the Pharaohs*. Rochester, VT: Bear & Company.

Edwards, I. E. S. 1993. *The Pyramids of Egypt*. London: Penguin Books Ltd.

Effat, Mahmoud, Fathi Saleh, and Robert Gribbs. 2000. "On the Discovery of the Ancient Egyptian Musical Scale." EgyptSound.fr website. Accessed April 28, 2022.

El-Aref, Nevine. 2017. "Void within Great Pyramid Of Giza 'not a new discovery': Zahi Hawass" Ahram Online website, November 2, 2017.

Elfouly, Gamal. 2012. *The Great Pyramid System: The Blue Light*. CreateSpace Independent Publishing.

Ellis, Ralph and Mark Foster. 2000. "The Enigma of Mamun's Tunnel." *Atlantis Rising* 25, 2000.

Energy Market Authority. n.d. "Singapore Energy Statistics, Chapter 2: Energy Transformation." Energy Market Authority website.

Fakhry, Ahmed. 1969. *The Pyramids*. Chicago: University of Chicago Press.

Finholt, A. E., A. C. Bond Jr., and H. I. Schlesinger. 1947. "Lithium Aluminum Hydride, Aluminum Hydride and Lithium Gallium Hydride, and Some of their Applications in Organic and Inorganic Chemistry." *J. American Chemical Society* 69 (5): 1199–1203.

Florinsky, Igor V. 2016. "Earthquake Lights in Legends of the Greek Orthodoxy." *Mediterranean Archaeology and Archaeometry* Vol. 16, no. 1 (January 2016): 159–168.

Freund, Friedemann. 2000. "Time-Resolved Study of Charge Generation and Propagation in Igneous Rocks." *Journal of Geophysical Research* 105, no. B5 (May 10, 2000): 11001–11019.

———. 2003. "Rocks That Crackle and Sparkle and Glow: Strange Pre-Earthquake Phenomena." *Journal of Scientific Exploration* Vol. 17, no. 1 (2003): 37–71.

———. 2009. "Pre-Earthquake Signals Introduction to Basic Solid State Processes." *Contemporary Physics* (August 10, 2009): 36.

———. 2013. "Nature of Pre-Earthquake Phenomena and Their Effects on Living Organisms." *National Library of Medicine: Center for Biotechnology Information*. Published online 2013 Jun 6.

———. 2014. "Rocks that Crackle and Glow: Predicting Earthquakes." Filmed at the 33rd Annual Conference of the Society for Scientific Exploration,

June 6–7, 2014, San Francisco, California. Society for Scientific Exploration YouTube channel. November 21, 2019.

———. 2016. "Using Semiconductor Physics to Forecast Earthquakes." TEDx talk, Christchurch, New Zealand, December 12, 2016. TedX Talks YouTube Channel. Accessed April 13, 2022.

Freund, Friedemann, G. Ouillon, J. Scoville, and D. Sornette. 2021. "Earthquake Precursors in the Light of Peroxy Defects Theory: Critical Review of Systematic Observations." *European Physical Journal* (2021). Special Topics 230, 7–46 (2021).

Freund, F., V. Stolc. 2013. "Nature of Pre-Earthquake Phenomena and their Effects on Living Organisms." *Animals (Basel)* 3, no. 2 (June 6, 2013): 513–31.

Ghoneim, M. Zakaria. 1956. *The Lost Pyramid*. New York: Rinehart & Company.

Gough, Evan. 2022. "Archeologists To Scan The Great Pyramid Of Giza With Cosmic Rays." Universe Today website. February 28, 2022. Accessed May 8, 2023

Hawass, Zahi. 2014. "The So-Called Secret Doors Inside Khufu's Pyramid." *Ministry of Antiquities* published in *CASAE* 43 (2014): 51–68.

Hays, W. W. (ed.). 1981. *Facing Geologic and Hydrologic Hazards—Earth Science Considerations*. US Geological Survey, Professional Paper 1240-B.

History's Mysteries. 1995. "Secrets of the Pyramids and the Sphinx." Maximum Truthseeker YouTube Channel. Accessed April 11, 2022.

Hooper, Rowan. 2011. "First Images from Great Pyramid's Chamber of Secrets." New Scientist website. May 25, 2011. Accessed April 27, 2022.

HyperPhysics. n.d. "Reflection of Sound." HyperPhysics website.

Karpenko, V. and J.A. Norris. 2002. "Vitriol in the History of Chemistry." *Chemické Listy* 96 (2002): 997–1005.

Kato, Mamoru. 2007. "Why Light Flashes—Light Emissions before and during Earthquakes: Presenting the Witnesses and Scientific Experiments." Paper presented to a conference held at the Earthquake and Volcano Research Center, Graduate School of Environmental Studies, Nagoya University, Japan, January 26, 2007.

Kingsland, William. 1996. *Great Pyramid in Fact and Fiction*. Kessinger Publishing, LLC. Facsimile edition, April 1, 1996.

Kunkel, Edward J. 1962. *The Pharaoh's Pump*. Self-published. Reprints from Peg's Print Shop; rev. and enlarged edition, January 1, 1973.

Lee, W. H. K., M. Ando, and W. H. Kautz. 1976. "A Summary of the Literature on Unusual Animal Behavior Before Earthquakes." In *Abnormal Animal Behavior Prior to Earthquakes*, J.F. Evernden (ed.), National Earthquake Hazards Reduction Program, USGS, Menlo Park, CA, 23–24 September 1976, 15–54.

Lehner, Mark. 1997. *The Complete Pyramids. Solving the Ancient Mysteries.* London: Thames & Hudson.

Lehner, Mark and Zahi Hawass. 2017. *Giza and the Pyramids: The Definitive History.* Chicago: University of Chicago Press.

Maragioglio, Vito and Celeste Rinaldi. 1965. *L'architettura delle piramidi Menfite. Parte IV (2 vols).* English transl. by J. A. Zanini Jellis and V. Maragioglio. Tavole. Rapallo: Tipografia Canessa: Rapallo, 1965.

McGovern, P. E. 1997. "Wine of Egypt's Golden Age: An Archaeochemical Perspective." *J. Egyptian Archaeology* 83 (1997): 69–108.

Morishima, Kunihiro, Mitsuaki Kuno, and Mehdi Tayoubi. 2017. "Discovery of a Big Void in Khufu's Pyramid by Observation of Cosmic-Ray Muons." *Nature.* 552 (November 2, 2017): 386–390.

National Geographic. 2002. "Into The Great Pyramid." The Last Pyramid Builder YouTube Channel. February 23, 2020. Accessed April 11, 2022.

New Madrid. n.d. "Strange Happenings during the Earthquakes." New Madrid, Missouri website. Accessed June 24, 2022.

Noone, Richard W. 1986. *5/5/2000: Ice: The Ultimate Disaster.* New York. Harmony Books.

ODNI. 2021. *Preliminary Assessment: Unidentified Aerial Phenomena.* Office of the Director of National Intelligence website. June 25, 2021. Accessed 2/17/2022.

Pacific Coastal and Marine Science Center. 2018. "Tsunami Generation from the 2004 M=9.1 Sumatra-Andaman Earthquake." US Geological Survey website. October 8, 2018. Accessed April 11, 2022.

Petrie, William Flinders. 1883. *Pyramids and Temples of Gizeh.* London, England: Field & Tuer. Reprinted 1990. London, England: Histories and Mysteries of Man.

Piroddi, L., G. Ranieri, F. Freund, and A. Trogu. "Geology, Tectonics and Topography Underlined by L'Aquila Earthquake TIR Precursors." *Geophysics Journal International* 197 (2014): 1532–1536.

Pond, Dale and Walter Baumgartner. 1995. *Nikola Tesla's Earthquake Machine.* Santa Fe, NM: The Message Company.

Qvist, Mark. 2023. "Abstractions Set In Granite." March 18, 2023. Unsigned.Io Website. Accessed July 6, 2023.

Roadside America website. n.d. "Cursed Pyramid Ruins." Accessed June 24, 2022.

Rubinstein, William D. 2008. *Shadow Pasts: History's Mysteries.* Harlow, UK: Pearson Education Limited.

Sagan, Carl. 1996. *The Demon-Haunted World: Science as a Candle in the Dark.* New York: Ballantine Books.

Schoch, Robert M. 1992. "How Old is the Sphinx?" Abstracts for the Annual Meeting of the American Association for the Advancement of Science, Chicago (1992): 202.

Schoch, Robert M. 2010. "The Mystery of Göbekli Tepe and Its Message to Us." New Dawn Magazine 122 (Sept–Oct 2010).

Schoch, Robert. 2018. "The Big Void." *Atlantis Rising* 128 (June–September, 2018).

ScoopEmpire. 2017. "Meet Aliaa Ismail, Egypt's First Female Egyptologist." November 11, 2017. Scoop Empire Website. Accessed May 23, 2022.

Schröder, G. 1957. "Die pharmazeutisch-chemischen Produkte deutscher Apotheken im Zeitalter der Chemiatrie." Technischen Hochschule Braunschweig, Bremen (1957): 61.

Selin, Shannon. 2017. "Napoleon at the Pyramids: Myth Versus Fact." Shannon Selin website.

Seyzfadeh, Manu, 2021. *Under the Sphinx: the Search for the Hieroglyphic Key to the Real Hall of Records.* Englewood, CO. Hugo House Publishers.

Sibson, Matt. 2020. "The Robot, The Dentist and the Pyramid: Ancient Egypt Documentary (2020)." Ancient Architects YouTube channel, February 19, 2020. Last accessed April 27, 2022.

Sibson, Matt. 2021. "First Look Inside the Great Pyramid Queen's Chamber Northern Shaft." Ancient Architects YouTube channel, July 28, 2021. Last accessed April 26, 2022.

Smyth, Charles Piazzi. 1978. *The Great Pyramid: Its Secrets and Mysteries Revealed.* New York, NY. Bell Publishing Company. Originally published in 1880 as *Our Inheritance in the Great Pyramid.*

Strutzenberger, J. and J. Faderl. 1998. "Solidification and Spangle Formation of Hot-Dip-Galvanized Zinc Coatings." *Metallurgical and Materials Transactions* A 29 (February 1998): 631–646.

Sutton, M. A., J. W. Erisman, F. Dentener, and D. Möller. 2008. "Ammonia in the Environment: From Ancient Times to the Present." *Environmental Pollution* 156 (2008): 583–604.

Takeuchi, Akihiro, O'mer Aydan, Keizo Sayanagi, and Toshiyasu Nagao. 2011. "Generation of Electromotive Force Due to Non-Uniform Compression of Igneous Rock and its Mechanism." Tokai University Marine Research Institute Research, Report No. 32 (2011): 45–51.

Taylor, John. (1859) 2014. *The Great Pyramid: Why Was It Built? And Who Built It?* Cambridge, UK: Cambridge University Press.

The Engineering Toolbox website. 2022. "Solids and Metals—Speed of Sound." Accessed May 25, 2022.

Tesla, Nikola and the Secret Libraries. 2016. *Nikola Tesla Quotes . . . Vol.3: Motivational & Inspirational Life Quotes by Nikola Tesla.* CreateSpace Independent Publishing Platform.

Thériault, R., F. St-Laurent, F. Freund, and J. S. Derr. "Prevalence of Earthquake Lights Associated with Rift Environments." *Seismological Research Letters* (January 2014).

Tompkins, Peter. 1971. *Secrets of the Great Pyramid.* New York: Harper & Row.

Tributsch, H. 1982. *When the Snakes Awake.* Cambridge: MIT Press.

US Energy Information Administration. n.d. "What is U.S. Electricity Generation by Energy Source?" US Energy Information Administration website. Accessed March 4, 2022.

Uyeshima, M. (2007). "EM Monitoring of Crustal Processes Including the Use of the Network-MT Observations." *Surveys in Geophysics* (May 2007).

Vitruvius. 1914. *The Ten Books on Architecture.* Translated by Morris Hicky Morgan and Albert A. Howard. Cambridge, MA: Harvard University Press.

White, George S. (1836) 2012. *Memoir of Samuel Slater: The Father of American Manufacturers.* London: Forgotten Books.

Wikipedia. 2011. "Electricity Sector in the United Kingdom." Accessed April 11, 2022. Last edited April 6, 2022.

Wikipedia. 2022. "2009 L'Aquila earthquake." Accessed April 11, 2022. Last edited March 16, 2022.

Zeigarnik, Vladimir and Victor Novikov. 2014. "Electromagnetic Earthquake Triggering Phenomena: State-of-the-Art Research and Future Developments." Geophysical Research Abstracts Vol. 16, EGU2014-16221, EGU General Assembly 2014, held April 27–May 2, 2014, in Vienna, Austria.

INDEX

Numbers in *italics* preceded by *pl.* indicate color insert plate numbers.